ATF Science and Theology Series: 3

CW00363231

Habitat of Grace
Biology, Christianity and the global environmental crisis

Carolyn M King

The ATF Science and Theology Series is a publication of the Australian Theological Forum. Each volume is a collection of essays by one or a number of authors in the area of science and theology. The Series addresses particular themes in the nexus between the two disciplines, and draws upon the expertise of both scientists and theologians.

Series Editor
Mark Wm Worthing

Australian Theological Forum, Adelaide

ATF Science and Theology Series
Series Editor: Mark Wm Worthing

1: *God, Life, Intelligence and the Universe,* edited by
Terence J Kelly SJ and Hilary D Regan, 2002

2. *Interdisciplinary Perspectives on Cosmology and
Biological Evolution,* edited by Hilary D Regan and Mark
Wm Worthing, 2002

Habitat of Grace
Biology, Christianity, and the global environmental crisis

Carolyn M King

Senior Lecturer in Biological Sciences
University of Waikato, Hamilton, New Zealand

First published January 2002

National Library of Australia
Cataloguing-in-Publication data

King, C.M. (Carolyn Mary)
Habitat of Grace: Biology, Christianity, and the global environmental crisis

Bibliography
ISBN 0 958639981

1. Religion and science. 2. Human ecology – Religious aspects – Christianity. I. Australian Theological Forum. II. Title (Series: ATF series; 3).

215

Published by
Australian Theological Forum
P O Box 504
Hindmarsh
SA 5007
Australia
ABN 68 314 074 034
www.atf.org.au

Printed by Openbook Publishers, Adelaide, Australia

Contents

Contents *cont'd*

Contents *cont'd*

List of Figures and Tables

Acknowledgments

To the very many people who have helped me in various ways to complete this book, my heartfelt thanks.

It started as an M Phil thesis for the University of Waikato, and over the next five years grew into a doctorate. For initial encouragement and support, I thank: Bishop David Moxon (Diocese of Waikato), Prof Warwick Silvester (University of Waikato), Prof Emeritus John Morton (Auckland) and members of the two EFM groups with whom I studied from 1989 to 1996 (led by Cherry Dingemans and Elizabeth Hopper). My title, *Habitat of Grace*, is borrowed from Morton's book *Christ, Creation and the Environment* (1989). For the necessary study grant: St John's College Trust, Auckland; for hospitality during a year's study leave I took back at my *alma mater*: The Master (Dr Richard Repp) and Fellows of St Cross College, Oxford. For supervision, discussions and advice: Dr Douglas Pratt (Waikato), Dr Stephen May (Auckland), Dr Arthur Peacocke and Dr Kevin Sharpe (Oxford). For consultations on particular issues: Rev. Kelvin Wright (Hamilton), Prof Ian Green (Queens University, Belfast), Christopher Hall and Prof Keith Ward (Oxford), Margaret Jeffrey and Dr Roger Williamson (Church House, London), Prof R J Berry (UCL London); Jan Richmond (Otaki). For drawing Fig 4: Dr R Brockie (Wellington).

I have of course fully referenced all ideas that I have knowingly quoted from others, and my debt to them is huge, as the bibliography shows. Statements or concepts that are not referenced are mostly my own. But it is in the nature of life in the academic community that a casual conversation or a passing comment from someone else may lie dormant in the back of the mind for some time, and only much later emerge as what might appear to be one's own original concept. I would be unusual among scholars if that process has not also contributed to this work. To those unknown contributors, I also offer my thanks.

For reading the manuscript, as a whole or in parts, and then providing many perceptive critical comments: Drs Pratt, Peacocke and Sharpe (twice each), Prof John Morton (Auckland), Bishop David Moxon, Dean John Fairbrother, Prof Warwick Silvester, Dr Al Gunn and

Ken Ayers (Hamilton). For conducting the viva and offering many suggestions for further improvement: Prof Berry and Prof Donovan (Massey University). Between them all these friends and colleagues saved me from any errors; as usual, those that remain are my own.

The design of a book addressed to a mixed audience, as this one is, required a difficult decision on how much background material to include. The science-religion debate is inhabited by two sorts of people: theologians exploring science, and scientists exploring theology. I have met far more of the first than of the second, and have also observed that their efforts to understand the arguments may be hampered if scientists fail to explain important background concepts that all scientists normally take for granted. Since this book is intended to benefit ordinary Christians, I have added an Appendix describing the workings of natural selection, in order to make it comprehensible to readers who have no specialist biological background.

Parts of this text were first published in theological journals (King 1997; 2001), and I acknowledge with thanks the permission given by their Editors to reproduce this material.

1. Science, Religion and Environment

1.1 Setting the scene

In 1990, the Worldwatch Institute in Washington estimated that humankind has about forty years to make the transition to 'an environmentally stable society'. If we have not succeeded by then, it concluded, 'environmental deterioration and economic decline are likely to be feeding on each other, pulling us into a downward spiral of social disintegration' (Brown and others 1990: 174). Worldwatch is no millenarian cult, but a sober and careful organisation whose annual summaries of world affairs have become the planet's unofficial environmental health reports. Its pronouncements are cautiously worded, influential, and worth attending to, even if the timing is hard to predict. ('Things drag', so, 'if Worldwatch says forty years we probably have seventy', says Randy Hayes, Rainforest Action Network, quoted by Athanasiou, 1996: 60).

At about the same time, the potential contribution of the world's religious communities to dealing with the environmental crisis was made explicit by a group of thirty-four leading scientists, including the atheist cosmologist Carl Sagan and the Marxist palaeontologist S J Gould. In 1990 they signed an Open Letter to the Religious Community seeking to enlist the help of people of faith in addressing the environmental crisis. The letter points out that problems as huge as the contemporary threats to planetary health must be recognised as having 'a religious as well as a scientific dimension . . . efforts to safeguard and cherish the environment need to be infused with a vision of the sacred' (Baker 1996; Rasmussen 1996: 183).

In 1992 came the World's Scientists' Warning to Humanity, signed by one thousand five-hundred and seventy-five distinguished scientists, including more than half of all living Nobel laureates (Ehrlich and Ehrlich 1996: 242). All these authoritative voices agree that the survival of western civilisation will be at stake in the forseeable future, certainly in the lifetimes of our children. Like the Open Letter to the Religious Community, the World's Scientists' Warning concluded with an appeal for help from the leaders of the global communities of businesses, industries and religions, yet the response so far has been minimal. Even the majority of their fellow-scientists have ignored the World's Scientists' Warning and, instead of working hard to inform policy-makers and public of the technical and social consequences of current demographic and ecological forecasts, continue to devote themselves to ever-more sophisticated analyses of trivial questions (Ehrlich 1997).

Various other watchdog organisations and influential voices such as Herman Daly, David Suzuki and the late Jacques Cousteau have called the 1990s the 'crucial decade', or 'the turning point for human civilisation'. They all agree that we must realise what is happening and start doing something decisive if we want to avoid witnessing western civilisation sink into an uncontrollable decline (Athanasiou 1996: 57). Or, 'most of the great environmental struggles will either be won or lost in the 1990s. By the next century it will be too late' (Thomas Lovejoy, quoted by McDonagh 1994: 145).

The data upon which predictions of future trouble rest are too diverse to be documented in full here, are not original to me and are (in outline at least) not in dispute. The argument I aim to present takes them as read and proceeds on from there. However, it is useful to have a reminder of the main facts, so I here reproduce much of a succinct summary provided, with references, by Jacobs (1996: 15-17).

Global climate change. The scientific consensus on global climate change is represented by the findings of the Intergovernmental Panel on Climate Change (IPCC), comprising several hundred of the world's leading atmospheric scientists. The IPCC has confirmed the very strong probability that, as a result of emissions of greenhouse gases—particularly carbon dioxide, methane and CFCs—global mean temperatures will rise at a rate of around $0.3°C$ per decade—faster than at any time in the last ten thousand years. Mean temperatures are likely to rise by about $1°C$ by 2025 and $3°C$ by 2099. Sea levels will rise by about 65 cm by 2100. There has already been a noticeable rise in global temperature over the last few decades.

Ozone depletion. The release of CFCs into the upper atmosphere continues to deplete the ozone layer that normally shields earthly life from ultraviolet radiation. A 'hole' in the ozone layer over the Antarctic has been documented every year since 1982, and in 1995 a similar 'hole' appeared over the Arctic as well. The consequences are expected to include increased incidence of cataracts and skin cancers plus many so-far unknown effects on ecological systems. If all countries comply with their obligations under the 1987 Montreal Protocol (an international agreement to phase out the production and use of CFCs), the damaging trend could be arrested after about 2002, but full repair, if possible at all, cannot be expected before 2050.

Deforestation. During the 1980s alone, about eight per cent of the world's remaining tropical forest was lost, amounting to an area almost three times the size of France. The main causes, varying in different countries, are commercial logging, beef ranching and population resettlement. Tropical forests are home to many indigenous peoples and endangered species; they absorb CO_2 and so help to arrest the greenhouse effect; and they are the potentially sustainable source of many drugs and industrial products such as rubber, plant oils and resins.

Biodiversity. The continuing loss of global biodiversity, due mainly to habitat destruction (especially of tropical forest) and the translocation of species around the world by human agency, is now proceeding much faster than the 'normal' natural rate (ie, the average for times other than during the mass extinctions triggered by large cosmic impacts). Projections from current trends suggest that between one per cent and eleven per cent of the world's species will have become extinct every decade between 1975 and 2015. At present, about twelve per cent of all mammalian species and about eleven per cent of all bird species are listed as threatened, and a disproportionate number of the list of threatened birds are endemic to New Zealand.

Fisheries. The UN's Food and Agriculture Organisation reports that catches exceed maximum* sustainable yield[1] in four of seventeen important marine fisheries, so stock reductions must already be in progress. Catches are declining in a further six areas. Most other traditional fish stocks have reached 'full exploitation', meaning that intensified effort cannot increase the catch, and new technology aimed

---

1. Words with asterisks are defined in the glossary.

at doing so will inevitably induce reductions in these stocks as well.

Water. Scarcity of fresh water is already an increasing problem in twenty-six countries, home to two hundred and thirty million people. 'Water scarcity' is defined as annual supplies of <1000 m^3 per person. Now, in the early 2000s, a third of the total population of Africa is living in water-scarce countries. Declining water tables, indicating unsustainable use, are now evident in parts of China, India, Mexico, western US, North Africa and the Middle East. Some seventy per cent of the world's drylands are suffering desertification, and about one and half million hectares of agricultural land are lost every year from salinisation. In most of Africa, per capita food production has been static or declining for the last decade even as population continues to increase.

Acidification. Pollution due to emissions of sulphur, nitrogen oxide and heavy metals is now affecting forests throughout the world. For example, acid rain affects twenty-two per cent of European forests, and fourteen per cent of the land area of China (and since China plans to increase the extraction of sulphur-rich coal by thirty-five per cent in the near future, the latter total is bound to increase). Air and water pollution and the production of hazardous wastes are increasing rapidly in almost all industrialising countries.

Pesticides. Chemical fertilisers, irrigation and the development of high-yield crop varieties have increased agricultural productivity, but also made crops more vulnerable to attack by pests and consequently increased their dependency on regular use of pesticides. But pests quickly become resistant, setting up a 'treadmill' of ever-increasing pesticide use, which itself creates serious problems for the health of humans and natural environments. In the fifty years since the regular use of pesticides became widespread, the percentage of crop loss due to pest damage has not measurably declined.

These and many more facts lay behind the sobering words of Boutros Boutros-Ghali, then Secretary-General of the UN, in his opening address to the UNCED* conference in Rio (June 1992): 'The time of the finite world has come, in which we are under house arrest . . . Nature no longer exists in the classic sense of the term' (Granberg-Michaelson 1992: 7).

People vary in their reactions to this litany of depressing data. Some become fearful and apathetic; others take refuge in ignorance or escapism (such as, 'There is no need to do anything, God is in charge of all history'). Furthermore, all these undoubted threats to the natural world are only the start of it: there are even more problems besieging the human social environment. Poverty, injustice and environmental degradation go together, since the poor almost always live in the worst environments. Exhortations by green activists to get people to vote on green issues tend to fall on deaf ears, or underline the feelings of helplessness among the sensitive. New Zealand developed the first green political party in the world (Values, established in 1972), but was slow to learn from its history. After more than twenty years, Rainbow (1993) was despairing that there was still no clear green vision, and little on the political scene had changed. Maybe, since he wrote, the tide could have turned: the desired change of scene in New Zealand apparently arrived after the 1999 General Election, when a new Green Party was suddenly catapulted into the most influential position possible, the balance of power between the two traditional parties and their allies. The long term benefits of how they will use that influence remain to be seen.

Part of the reason that green politics was so difficult for so long is that, in a traditional first-past-the-post election, minor parties have seldom been able to gain enough local influence to win more than the occasional seat. Apprehension among ordinary people prepared to think about what is going on, yet unable to influence the direction of governmental decisions, generated a sense of impotence. The forces of environmental change seem to be so rapid and uncontrollable, so hugely threatening to both the natural and human worlds; and yet at the same time our political systems have simply not been addressing them—as if politicians were not living in the real world (Jacobs 1996: 2). However, in 1996 New Zealand's system of voting was changed to Mixed-Member Proportional Representation (MMP). Under MMP, the proportion of the nation-wide party vote gained (above a threshold of five per cent) by each party determines the allocation of seats in the one hundred and twenty member House of Representatives. At the 1999 General Election this system gave the Greens seven seats, an effective caucus, and a real reason for optimism among those who have long hoped to see environmental concerns taken seriously in New Zealand political decisions.

On the other hand, the problem is far more complex than party politics. Individual politicians and scientists are as caught up in indecision as anyone, because at root, what we call the 'environmental crisis' is a sign of the cultural, and ultimately spiritual, failure of the whole of western society and most of its leaders. We all have to find a compromise between our personal convictions and the political and economic realities that govern us. None of us can live outside the prevailing social matrix—even the most committed environmentalists still fly around the world to meetings discussing what we might do about the environmental crisis which is in part accelerated by massive use of jet fuel (Oelschlaeger 1994: 3). The ultimate problem of the modern world, said Vaclav Havel, then Czech President, is 'a lost integrity' (quoted by Rasmussen, 1996: 17-19). We are all stuck together in the awkward space between a past we once trusted and a future we have not begun to understand. Yet the mainstream religions are generally seen as having nothing to say, as obsolete distractions (Rasmussen 1996: 10).

1.2 Rationale

I embarked on this study in order to resolve a contradiction of which I have been aware for many years. If, for Christians, faith in a loving God is the ground of all meaning, as theologians assert and as I have found throughout a turbulent life, why does it have nothing to say about all this, the most urgent crisis of meaning in human history? Despite its ambiguities and chequered past, Christianity is a radically incarnational*, historical faith, anchored to the material world of time and space and insisting that creation and redemption are matters of direct, earthy experience in the present world as we know it (Houghton 1997). Of all the mainstream religions, surely Christianity ought to be able to say something useful about 'lost integrity', and to avoid isolating discussions of conservation or environmental problems from the related questions of social justice and of the advance of science, or from the great religious questions that undergird them all. But I was frustrated by the apparent lack of any sign, or at least any that was visible to me in my pew week after week, that any church leaders ever thought about the theological dimensions of these vital issues.

Part of this book was written during a year's study leave which I spent at St Cross College, Oxford—the same College from which I had taken a doctorate in zoology and immediately took up a research appointment in New Zealand in 1971. Returning to the city after twenty-five years brought many poignant experiences, among which the most astonishing and the most disturbing was the regular sight of beggars in the street, which I had never seen in my student days. Walking the scant one hundred metres between College and the Theology Faculty Library was often an expensive business. I constantly struggled to reconcile the security, wealth and civility inside the colleges with the insecurity, poverty and degradation outside them; the immense gap now visible between the haves and the have-nots with my memories of a more egalitarian British society in the 1960s; and the urge to follow the teachings of Christ about charity to the poor with the fact, or so I thought, that British taxpayers were already funding a welfare state supposed to meet the needs of the disadvantaged before they fell into undignified destitution on the street. Such encounters were as significant as regular lectures in making me think about the deeper questions underlying my subject.

The international connections between beggars in an Oxford street and the economic and religious dimensions of conservation on New Zealand are neatly summarised by Ambler (1990: 7-9) in his short, dense book *Global Theology*:

> When so much of our life in the West is bound up with the hope of an ever-increasing expansion, how can we envisage life without it . . . how can we believe . . . that the civilisation that has given us modern science, democracy, individual rights and freedoms, and unparalleled wealth and leisure should be too unstable to last—and more than that, that it should be responsible for its own downfall? . . . what begins as a material crisis . . . becomes eventually a spiritual crisis

. . . the material threats posed are directly or indirectly the result of our pursuit of material security. This paradox questions some of the deepest assumptions on which our lives are based . . . questions for which no answer can be found in modern science or in the practical wisdom of our future-oriented age . . . We must find a discourse in which we can communicate both our global concerns and our spiritual insights, our questions and our answers. We need something like global theology.

Theology may not be the best word: it has perhaps too many associations with abstract theorising and Christian apologetics . . . [but] it is the only word we have for the job in hand, namely a serious intellectual enquiry into matters of spiritual concern. *When theology is doing its job and not being diverted into defensive or abstruse argumentation, it is concerned with the central issues of the day and of human life itself . . . [and] with those responses to life in the past or today which seem to resolve the central problems of life through . . . confidence in the ultimate reality on which our life is based* [my emphasis]. We can call those responses 'faith' as a kind of shorthand, but faith should then be distinguished from religious dogmas which may or may not give expression to faith. Theology could then be described as an attempt to understand the meaning of faith . . . in different and changing circumstances. Global theology would be a search for the meaning of faith in our new global situation . . . If we hope to respond adequately to the present world crisis we need both to cross the boundaries of old religions and ideologies, and to dig deeply into the basic religious questions about human identity and security.

To anyone outside the churches, Ambler's view, that one can go directly from the environmental crisis as the problem to theology as the answer, may seem like a huge and unjustified leap in logic. The sort of theology that Ambler has in mind would have to find its place in a real world dominated by other systems of thought that do not, to put it mildly, regard theology as having anything vital to contribute. It is therefore necessary to examine Ambler's questions more closely in the light of recent advances in science* (especially ethology*, evolutionary psychology* and ecology), economics and politics, before attempting to put them into theological context.

Of course, that approach does not imply that theology must be judged in terms of science, or vice versa, although many are ready to take that view, eg Michael Cavanaugh (1996). It is the dialogue between them that might help, which in turn requires a deliberate broadening of vision—the opposite of the methodological reductionism* typical of the scientific view taken alone. Science can only propose rational arguments about life; our duty is to consider their significance in the light of our non-rational experience of Christian sacramental spirituality. The field is very large, and I agree with Donald Campbell that the problem of trying to cover more than one specialised academic discipline is that the attempt, though necessary to gain a wider view, also requires an author to be 'willing to be incompetent in a number of fields at once' (quoted by Ruse 1986: xiii). Becoming familiar with that risk throughout my work proved to be the best possible preparation for appreciating the only book I have read that seriously challenges Campbell's statement, E O Wilson's *Consilence: the Unity of Knowledge* (1998).

1.3 Thinking about science and theology

There are very many possible definitions of science, but they all involve concepts of speculation, experimentation, evidence, falsification, scepticism and constant revision under peer review. Because it is essentially practical and experimental, many people think of science as 'organised common sense'. Somewhat fewer prefer to emphasise the provisional character of all theories by calling it 'a series of successive approximations to the truth'. Almost everyone assumes that science is useful, reasonable, and closely related to technology, because it is supposed to be the basis of technological advance, or rather, that technological advances depend on scientific understanding of the underlying processes. Hence 'Science and Technology' are usually bracketed together, in Departmental titles, grant application procedures and ministerial portfolios.

Lewis Wolpert, in *The Unnatural Nature of Science* (1992), argues that the label 'organised common sense' is properly applied only to technology, and that technology alone produces all the 'useful' aspects of research—including almost everything that most of us think of as science, from computer engineering to clinical medicine. Science itself, Wolpert says, is a very different matter from technology. It is concerned with abstract speculations about the nature and behaviour of matter, and with fundamental explanations of the processes governing the material world. It is not itself directly useful, it is often counterintuitive, and is usually irrelevant to the short-term decisions of daily life. Technological advance does not need science defined in this sense, and indeed is regularly made with no understanding of the underlying processes at all—for example, natural selection has produced marvellous technology in animals without it, and cultural selection has done the same for every independent human society. The same principles behind most technological inventions of the twentieth century have been exploited by animals for millions of years (the jet engine by squids, the heat exchanger by arctic mammals, electricity by certain fish, hydraulic power by worms, and so on); early hominids made stone tools and spear throwers without any knowledge of lithology or physics; herbal remedies, the inventions of the wheel and the bicycle owe nothing to science; and good cooking needs no understanding of chemistry. All these skills were developed by trial and error and cumulative natural or cultural selection, and handed down the generations.

By contrast, Wolpert pointed out, pure science requires a certain way of rigorous, often quantitative thinking that requires the abandonment of common sense and of its associated cultural authority. It takes an unusual person to be able to think as critically and freely as that, and such people are rare. One was Charles Darwin, who saw how the raw materials for evolution were supplied by the random variations which, though generally harmful to their owners, yet over the long term add up to precise, co-ordinated and beneficial adaptations. Darwin's insight is strongly counter-intuitive, and meets widespread incomprehension even today, despite the articulate explanations of many authors specifically aiming to combat this problem, such as Richard Dawkins in *Climbing Mount Improbable* (1996).

Not everyone would agree with Wolpert's definition of science. In *Taking Darwin Seriously*, Michael Ruse (1986) advocates a more evolutionary attitude to epistemology. He points out that scientific change is cumulative, as is evolution, though the analogy must be carefully qualified, since science is also directed and progressive as evolution is not. But in my opinion, they are both right. Wolpert's idea explains how individuals come up with original and insightful hypotheses (equivalent to mutations), while Ruse explains how the few hypotheses that survive the author's own self-criticism are then selected and developed by the scientific community.

There are as many different definitions of theology as there are of science. Theology was once known as 'the queen of sciences', although few would grant it that title now. But if Wolpert is right, theology could be closely analogous to pure science. If science comprises abstract speculation about the nature of matter, as distinct from technology defined as the practical management of life in the material world, then theology comprises abstract speculation about the nature of God, as distinct from religion defined as the practical management of life in the social and spiritual world.

Wolpert's interpretations of the relationship between science and technology suggest several other possible parallels for the relationship between theology and religion. First, religion is an essentially practical matter, strongly influenced by cultural authority, and for the vast majority of religious people it is conducted without conscious reference to the theological concepts behind it. Just as cooks do not need to understand chemistry in order to produce a good dinner, a believer does not need to understand any of the various theories of atonement in order to find peace and forgiveness in the sacrament of penance. In that respect, theology is no more directly useful to everyday life than pure science is, although of course both have profound hidden influence.

Second, Wolpert's claim that science is often counter-intuitive applies even more forcefully to theology. For example, the concept of unmerited grace is very difficult to grasp by human beings adapted to live in a society whose internal workings are governed largely by reciprocal altruism, which itself goes back to our simian ancestors (pp.76, 210). Current research in primatology is revealing the depth and influence of reciprocal altruism in our closest relatives, but also its flexibility and potential for community integration. In the light of our long evolutionary background, and on the assumption that human nature is at least partly explicable by it, we can all too easily understand why many people in the post-Christian world identify more with the prodigal son's older brother than with the prodigal himself, or more with the labourers who had worked in the vineyard since dawn than with those who came late but got the same wage. Both stories were told to convey a deep theological truth about the unconditional love of God, which sounded counter-intuitive and even offensive to the practical, common-sensical, religious people that first heard them. Likewise, people today still insist on keeping scores, despite centuries of Christian teaching. Most of Robert Farrar Capon's books, especially *The Parables of Grace* (1983) and *The Astonished Heart* (1996) are about why 'hell is full of forgiven sinners' who could never bring themselves to accept the gift of life for nothing.

Third, really rigorous and independent thinking is very difficult, and few people are capable of it. In pure science it requires a particular mindset, and a disciplined pre-occupation with apparently useless questions. Prayer also requires a particular mindset, and is a difficult, disciplined and apparently useless activity.

Fourth, any scientist who does come up with an original idea subjects it to experimental testing against the real world and ultimately to the judgement of the scientific community. Hence it is claimed that science is objective in a sense that theology can never be. Theologians may well attribute a profoundly original insight, in themselves or others, to a divine revelation, but ultimately that does not allow them to escape the need to achieve congruence with the real world, or to submit to the judgement of the religious community. The contemporary rejection, especially since the rise of feminism, of traditional doctrines on human nature based on Augustinian interpretations of the Fall seems to me to be effectively a religious equivalent to an unfavourable referee's report in the peer-review processes of science. What Barbour (1997:158) calls 'intersubjective testing' of belief—the long process of filtering, confirming and publicly validating our individual and communal experience—is much slower and less rigorous in religious than in scientific communities, but it is there.

In short, science is not organised common sense, and neither is theology. They have much more in common than appears at first sight: both can be distinguished from their respective practical aspects, and both are in constant need of revision, even though that is never easy. Wolpert concludes that 'Only a small amount of science is useful; the trouble is, we don't know which bit'.[2] In my view, the same applies to theology, and that parallel is at the same time the toughest challenge to, and the best hope for, the ancient traditions brought into question by the contemporary environmental crisis. Under these definitions, this book is mainly about the religious aspects of the environmental crisis and about the visionary faith and leadership that the churches must find in order to face it.

1.4 Faith, religion, and doctrine

There is an important distinction between the personal faith of individual believers, and the official structures (both the material buildings and the intellectual doctrines) of their particular religious tradition. This book is concerned with mainstream science and mainstream religion, and especially but not only with Anglicanism. I think it is possible to be, like Barbour (1997: 98), loyal to both science and faith, and yet also critical of both the reductionist and materialist interpretations of many scientists and the irrational and exclusivist interpretations of many religious dogmas. When I realised the significance of this distinction, in about 1989, I developed an appropriately biological metaphor to explain its significance to this study.

The Anglican Church*—the only one I know well enough to

---

2. Verbal comment made during a debate in Oxford in 1997.

comment on—is an extraordinary mixture of objective and subjective values, of matter and life, body and spirit, human and divine, religion and faith. All these are necessary before it can be fully itself. But they operate in different, overlapping domains of human experience. Challenges to deeply-held convictions belonging to one domain do not necessarily threaten the others. In the context of this book, it is very important to state clearly at the outset my conviction that it is possible to criticise religion and doctrine—and to change them—without damaging the faith that underlies them. The argument can be best explained with a metaphor involving a familiar animal, a hermit crab.

Like a crab, the church has an external skeleton, a visible and objectively definable container for the animal within. Ask ten people to define a crab, and their descriptions will be similar enough to instruct an eleventh person on how to recognise a crab in a rock pool. But the essence of a crab, the life of it, its essential crabbiness, so to speak, is not in its shell but inside, an invisible power in the soft parts that makes them pulse with blood and buzz with nervous messages. Ask the same ten people to define what makes a live crab alive, and they will probably produce quite a variety of definitions of life, even if they are scientists[3]. Maybe the only common conclusion will be that a dead crab cast up on the beach is still recognisably a crab, but it does not function as one.

A church building is still objectively definable as a religious institution *sensu* Wolpert, even when it is empty, but it cannot be a functional church until it is brought to life by the faith of its people. Conversely, there are places where groups of individuals meeting in each other's private homes blossom with vigorous faith, while the congregations supporting conventional religious observance in glorious surroundings wither away. Religion without faith is as dead as a cast-up crab; it is only faith, with or without religion, that can give life, and that life does not depend on any particular external form.

Every individual crab is the temporary expression of the crab-life

---

3. For two alternative current technical definitions, see Maynard Smith and Szathmary (1995: 17-18). A third, layman's interpretation might be: 'making something happen against the odds, and remembering how to do it' (Ursula Goodenough, President of IRAS*, comment during a lecture at the Star Island conference, 1996).

that it got from its parents and will pass on to its own offspring. The individual animals grow, breed and eventually die, but the genepool from which they came remains as alive, and as indefinable, as ever. The local population of crabs thrives for as long as their genes are passed in unbroken succession from one generation to the next. Moreover, the same properties of 'aliveness' that animate a live crab also pulsate through a huge diversity of other animals, from bacteria to elephants, each according to its own species pattern. In the same way, the faith that makes a church alive is passed on from one member of a congregation to another, and so long as the succession remains unbroken the group remains a functional church despite changes in individual membership.

Faith and genes are alike in that they have to be passed on from one individual to the next by direct contact, and in both, a break in the chain can cause a local extinction. The actual form of the body containing the life is not important; what matters is that that body works as an expression of the life inside it. There are hundreds of different types of crabs, and just as many different places where faith does its healing work.

The shells of crabs offer considerable protection from predators, but they cannot be invulnerable. There is one group of crabs, the hermit crabs, that has found an extra form of protection: they back themselves into a discarded mollusc shell (Fig 1). They have long naked tails, and they use them to hold onto an empty snail shell of the right size by extending the tail around the inside spiral. At the end of the tail there is a special appendage, with which a hermit crab can take a firm grip on the spiral. Thereafter it can move about, carrying the shell around on its back like a snail with legs. The snag about crab life in general is that, as the crab grows, its exoskeleton gets too tight, so it has to moult and grow a bigger one. The double snag for a growing hermit crab is that it not only has to grow a new exoskeleton, but it also then has to find a new shell of the right size, even if it has to be quite a different shape. The period between abandoning the old shell and finding a new one that fits is very dangerous, because that is when predators can catch the hermit crab with its vulnerable tail exposed.

Christianity in general is rather like a hermit crab. Experientially-defined faith gives Christianity life, which ultimately derives from the

Figure 1. A typical hermit crab (C M King, photo)

kerygma, the conviction that Jesus Christ is the Son of God and the saviour of the world. Christianity also has an exoskeleton, an objectively defined church or other religious institution in which that life is temporarily organised. But the church also has a soft tail, a point at which it is vulnerable to attack by its enemies in the outside world. To protect its tail, Christianity has developed since its earliest years a mass of traditions and doctrine (the *didache*), which are designed to defend its own members from damage and to answer the critics. Church buildings were themselves designed as visible representations of the current *didache*, and have changed as its doctrines developed. For example, engravings in the catacombs emphasised personal devotion to Christ in the Eucharist, but after Constantine, church art began to show him seated on a heavenly throne. Doctrines, like shells, that are effective for

16

the moment will not remain so indefinitely; live churches, like live crabs, cannot help growing out of them.

A hermit crab can live in a shell of various shapes so long as it is the right size. Likewise, faith can be expressed in various forms of *didache*, all equally valid provided they truthfully convey the central *kerygma* in a form that the current generation finds helpful. But just as a hermit crab is forced to move out of a shell that becomes unsuitable and find another one, faith that is driven out of a doctrine that is no longer tenable tends to turn up elsewhere.

The shell occupied by the catholic version of the Christian religion is particularly magnificent and has been growing for almost two thousand years, but it is now too heavy for the crab to move about in. Just as a species that was well adapted to a past environment may become endangered if that environment changes, so the post-Constantinian interpretation of the majesty of Christ is profoundly inappropriate for today's society. To me, the reason why the traditional churches have lost credibility in the contemporary world is not because the *kerygma* is no longer valid, but because its *didache* is no longer convincing —the crab is still alive but it needs a new shell, and quickly. Of course, it is during the period between shells that the crab is most vulnerable to attacks from outside; but the alternative is worse.

I am convinced that (1) it is possible to revise the *didache* without endangering the *kerygma*; and that (2) it is necessary to do this. The Angican Church recognises this, and has among its mission statements (p.34) one that envisages it entering into a fruitful partnership with the secular authorities who are already tackling the environmental crisis (King 2001). Other Christian and non-Christian religious traditions are doing the same. Without denying the organic connection between them, this book therefore makes a clear distinction between definable religious traditions and doctrine on the one hand, and experiential, indefinable, partly irrational (in the same sense that falling in love is to some extent irrational) faith and spirituality on the other.

Under these definitions, criticisms of, and suggested reinterpretations of doctrine are matters of rational judgement, and are not to be taken as criticisms of personal faith. Both are part of the religious experience, and there is ample evidence that deep religious commitment can be combined with critical reflection. The two pithiest summaries of this relationship are attributed to one authority from either side—a great saint, St Anselm ('theology is faith seeking understanding': McGrath 1994: 49) and a great scientist, Einstein ('science without religion is lame: religion without science is blind' : Wolpert 1992: 146-7). A recent survey confirmed that about forty per cent of practising scientists believe in a personal God and in life beyond death (Larson and Witham 1997)—about the same as in a previous survey taken in 1916.

1.5 Ways of approaching the contemporary dialogue between science and theology

Ordinary experience as a practising scientist soon shows that meaning and truth are culturally determined; that observations are never truly objective; and that all conclusions based on them, drawn from what one might hope is the unvarnished evidence of nature, are selective. Similar conclusions about the different sorts of meaning and truth conveyed by religious ritual are inevitable after a few decades of sitting in Anglican pews. A widely accepted fourfold classification of contemporary interactions between them is proposed by Barbour (1997).

Barbour's classification applies to science in general. In this work I refer mainly to biology, but with cross-connections to environmental ethics, politics and economics where necessary. I write as a biologist literate in Christian theology, aiming to explore the implications of my own discipline for the continuing development of Christian moral and practical theology. I avoid systematic and philosophical theology, as these terms are usually understood. I use the term 'religion' mainly in the Christian context, and other terms as defined in the Glossary.

1.5.1 Conflict

The most common image of the relationship between science and religion is that of warfare or conflict, fuelled by many repetitions of old stories such as the confrontation between Huxley and Wilberforce in Oxford (Lucas 1979), and in modern times relentlessly advocated by

Richard Dawkins and Peter Atkins, among others. The Oxford meeting and other historical events have been misreported and exaggerated, and in any case the diversity of views on both sides was far greater than the popular image allows, but these facts have not prevented the growth of two opposite forms of the conflict model, scientific materialism and Biblical literalism (Barbour 1997: 77). The mistake made by both is to assume that evolutionary theory is inherently atheistic, and they thereby perpetuate the false dilemma that people have to *choose* between science and religion. Barbour points out (84) that

> the whole controversy reflects the shortcomings of fragmented and specialised higher education. The training of scientists seldom includes any exposure to the history and philosophy of science or any reflection on the relation of science to society, to ethics or to religious thought. On the other hand, the clergy has little familiarity with science and is hesitant to discuss controversial subjects in the pulpit.

Underlying this problem is the even deeper question of what counts as a valid explanation (Poole 1994). The urgent need for the modern churches to address the matter of science education is taken up again later (p.148).

1.5.2 Independence

In their attempts to meet the challenge of science, defenders of traditional Christianity commonly argue that religion and science are compatible, because they are totally different. They do not compete because they address different questions—science explains how things are, religion explains why. Or, like Barth and his followers, they simply reject science altogether. But these are both unhelpful responses to the real challenge of science, and anyway they underestimate the capabilities, interests and aspirations of many scientists. Closer to the mark, and far less reassuring, is the totally opposite assertion of

Appleyard (1992: 83). He maintains that science and religion are completely incompatible, because

> Science [is] the lethally dispassionate search for truth in the world whatever its meaning might be; religion [is] the passionate search for meaning whatever the truth might be.

It is true that academics are used to thinking of truth as relative, experiental and conditional, whereas many ordinary Christians find it difficult to let go of the concept of truth as absolute, revealed and eternal. The difference between these views is the basis of Appleyard's astute but somewhat cynical observation on the different priorities of religious people and scientists. Attempts to harmonise them without guidance lead quickly to cognitive dissonance. For those unwilling to reject one or the other, the only possible response is to keep science and religion in separate boxes, and so retain the benefits of both without conflict. Such is the experience of countless youngsters brought up in Christian homes today.

In real life, science is not as objective, nor religion as subjective, as writers such as Appleyard like to claim (Barbour 1997: 93). The development of critical realism (p.25) makes it possible for members of both the scientific and religious communities to make cognitive claims about invisible realities. And we must do this, difficult as it may be, because as Barbour points out (1977: 89), if God is confined to the realm of the self, the natural order is deprived of any religious significance except as an impersonal stage for the drama of human existence. More significantly in the present context, this attitude offers the churches no hope of making any contribution to the environmental debate. If religion deals with God, and science with nature, who can say anything about the relationship between God and nature? So it is possible to argue on other grounds that the independence model is no more helpful to the science-religion debate than is the conflict model.

1.5.3 Dialogue

A somewhat diverse range of views can be grouped under this heading, which portray the relationships between science and religion as more constructive than under the conflict or independence models but not close enough to be thought of as integration. Some point to philosophical questions that are raised by science but cannot be answered by it, such as the apparently unique origin of science in western culture (p.129), or the 'fine-tuning' of the cosmological constants (listed byMcGrath 1998). Some list the methodological parallels, one form of which I illustrate on p.11. Some explore various forms of nature-centred spirituality, such as those of Matthew Fox, Rupert Sheldrake and Brian Swimme. The latter are characteristically much more holistic that most forms of mainstream (reductionist) science. Because there is much to criticise in reductionist science, especially reductionist biology (p.102), their views are interesting; but from the point of view of constructing a dialogue between church and state concerning the environmental crisis, Fox and Sheldrake are too controversial to be widely useful.

By far the most authoritative and most welcome participant in the dialogue is E O Wilson, himself usually regarded as one of the arch-reductionists. But his recent book *Consilience: The unity of knowledge* (Wilson 1998) is a modern treatise in the spirit of Thomas of Aquinas. Wilson and Aquinas both believe, for different reasons, that the world is orderly and potentially understandable in terms of a relatively few simple laws. Opposition to this view comes more often from the humanities than from the sciences, and Christian theologians can learn much from Wilson's reasoned responses to the vitriolic criticisms he has received from all sides. He is an atheist who leans 'towards deism but consider[s] its proof largely a problem in astrophysics' (*ibid*, 268). He concludes that religion will possess strength to the extent that it codifies and puts into enduring, poetic form the highest values of humanity consistent with empirical knowledge (296). Surely this statement is consistent with at least part of what rational Christians also believe.

1.5.4 Integration

Barbour (*ibid*, 98) identifies three strands of thought under this heading. Natural theology discusses to what extent it is possible to infer the character of God from evidence of design in nature. The theology of nature asks what input science may have in reformulating traditional, prescientific doctrines about creation and about the nature of humanity. Process theology aims to construct a coherent world view, an inclusive conceptual scheme which can interpret all events within a single system of metaphysics.

For many leading voices in this debate (Ian Barbour, Charles Birch, John Cobb and others) the development of process theology offers more hope of integration between religion and science than does traditional church teaching. Mesle (1993: 1) offers a simple definition: '"Process theology" is the name for an effort to make sense [ie, find truth and meaning], in the modern world, of the basic Christian faith that God is love . . . It requires that we rethink the nature of both God and the world'. Process theology can be criticised as downplaying the sovereign activity of God, both in creation (McGrath 1998: 46) and in redemption (Berry 1995), and such comments are helpful warnings against jettisoning established truth along with outmoded interpretations. However, this book is written from the conviction that it must be possible to see both scientific and religious truth as contributing parts of a greater whole, and that, however difficult and threatening the rethinking process may be, any progress we can make towards integrating them must surely help equip the churches to tackle the huge challenges raised by the environmental crisis.

1.6 Interpreting observations: models and paradigms in science and religion

Science and religion are both communal activities conducted under the pervasive but largely unconscious influence of a prevailing paradigm. Kühn (1970) defined a paradigm as a cluster of conceptual, metaphysical and methodological presuppositions held in common by those involved in a particular tradition, by which they interpret all observations of the world about them. We humans in general cannot make any sense of a new observation unless we can place it into the context of that prevailing paradigm, which can suggest from past experience some theory about what it might be. So C S Lewis describes the reaction of his hero Ransom on arriving on a totally unknown planet:

he knew nothing yet well enough to see it: you cannot
see things till you know roughly what they are (Lewis
1943).

Likewise, the first Aboriginals to see a sailing ship off their coast
ignored it, as they had no idea what it was and no experience from
which to guess the threat it presented to them (Willey 1979).

Observations of nature made in the days of the early Greek
philosophers such as Aristotle were filtered through minds filled with
contemporary theories about how the natural world worked. For
example, Aristotle believed that the behaviour of every natural object
was governed by its own nature–dogs bark, and rocks fall, because it is
their nature and purpose to do so. He never did any of the controlled
experiments that are the basis of modern science, because they could not
answer the questions he was interested in: he did not see the point in
interfering with the nature of the object being observed (Lindberg 1992:
53). The power of a ruling paradigm is well illustrated by the difficulty
that modern scientists have in understanding Aristotle's view.

Now our theories of nature are very different, and therefore the
questions we ask of nature and our interpretations of what we see are
also very different. The surprising thing is not that any of Aristotle's
interpretations were wrong, but that so many were right. For example,
Aristotle rejected the Platonic idea that the only true reality was to be
found in eternal forms, of which the whole material world as observed
by the senses—including all individual animals—are imperfect copies.
Aristotle's experience of watching and dissecting animals led him to
argue instead that individuals are real in and of themselves, and their
imperfections are part of their own character, not their failure to be
something else. On this point at least Aristotle sounds completely
modern, and very relevant to our contemporary understanding of the
key role of imperfections in the processes of evolution.

In both science and religion, the ruling paradigm defines what
questions may fruitfully be asked, what models of abstract reality might
be acceptable and what assumptions must be ruled out, how new data
will be sought and interpreted, and how old data can be understood in
new ways. An established paradigm is extremely resistant to

falsification, since discrepancies can, for a long time, be set aside as anomalies or interpreted through various auxiliary hypotheses which are much more easily modified than the central core theory (Murphy 1990). If the strain on the current interpretation becomes too great, there will be a sudden 'paradigm shift', during which some people suddenly 'see' the arguments from another point of view, and completely change their understanding of the matter. According to Hans Küng (1990), every major discontinuity in church history has been marked by a paradigm shift.

Where there is a choice between competing paradigms, the decision on which one to accept is personal and philosophical, since there are no objective rules to help. Hence, every paradigm shift in church history has left behind a group continuing to uphold the old ideas, and many of these have survived to the present day. In normal science such choices are rare, although I will argue later that recent work in primatology is now offering us two competing options on how to interpret the origins of human morality (p.92). On the wider scale, the conflict model (p.18) sees science and religion as offering competing paradigms by which the ordinary person may find meaning in life.

It is a well-worn truism, but no less true for that, that all data in science are theory-laden, and all experience in religion is concept-laden—there are no theory-free data and no uninterpreted experiences. The social and evolutionary experience of being human has long ago established unconscious preferences that influence what new knowledge we can take in. Together these constitute 'epigenetic rules' (Wilson 1998) or 'hermeneutical pre-knowledge' (Cavanaugh 1996: 250), which in turn determines what new explanations or proposals might be acceptable. So how do we answer Pilate's perceptive question, 'what is truth?' Barbour (1997: 109) lists three answers developed in western thought.

First, truth is correspondence with reality. This is an acceptable definition for common-sense situations, but fails at the door of any Wolperian research projects, or theological propositions, requiring the abandonment of common sense. Unfortunately, several concepts fundamental to resolving the environmental crisis appear quite contrary to common sense, eg that people in western democracies should vote for policies that would reduce their standard of living (p.42).

Second, a proposition is true if it is comprehensive and internally coherent. This criterion serves well until it meets another, competing but equally internally cohesive idea—such as during the argument over whether the universe was centred on the earth or on the sun. This book will offer examples of smaller-scale but similar meetings (p.112), since reality as we perceive it in nature and in the divine seems to be a lot more paradoxical and less logical than rationalists like to think.

Third, the pragmatic view is that a proposition is true if it works in practice—if it unifies disparate observations, and is fruitful in predicting new observations or finding solutions to puzzles. Barbour comments that the meaning of truth must be ultimate correspondence with reality, but since reality is often inaccessible to us, the criteria of truth must include all the above, taken together and welded into the system of thought, advocated by Barbour (1997: 118) and Peacocke (1995) and others, known as *critical realism**.

The basic assumption of critical realism in both science and theology is that existence precedes theorising. Both disciplines develop by the continual extension of the models that help us to visualise the invisible realities that surround us. Good models are metaphorical but not purely imaginary, since pre-existing realities always set limits to our speculations. They are neither rigidly defined nor merely temporary and dispensable, but fruitful and open-ended sources of ideas for cumulative modifications and extensions. The more we learn, the more we can adjust our models; contrariwise, models that are not adjusted to take account of new knowledge rapidly become irrelevant.

Influential individuals in both disciplines tend to become wedded to their models and equate modifications with personal criticism. This may impose humanly understandable but irritating delays in progress, but they cannot last for ever—as the saying goes, 'funeral by funeral, science advances'. The same applies to doctrinal developments, especially in the most conservative and oldest of institutions, such as the Roman Catholic Church. Therefore, a substantial part of this book is devoted to examining what advances might be made in Christian ideas about creation and human responsibility for it from current developments in evolutionary biology.

2. The Origin of Environmental Concern Among Christians

For Christians, there are three different aspects to the environmental crisis, all worrying. First is the challenge to our beliefs. There have been many attempts to understand the full implications of natural science for traditional theology, not all equally fruitful (Barbour 1997: 97). We do not have to deny the value and enduring truth of many older traditions (Morton 1989: 24), but at the same time we do have to allow for the fact that the relevant sciences (anthropology, evolutionary biology, ecology) are fast-moving international disciplines, and we must participate in the current debates, however scary they may appear. Christians have constantly to resist the temptation to 'domesticate' the debate, because, says John Reader,

> It is far easier and safer to try to contain the challenge of 'green' theology within existing boundaries than to be open to the possibility that what it really requires is a complete re-think of traditional Christian attitudes . . . most of the material that has been published so far has gone for the safe option, that of reinterpreting our existing language . . . what is needed is something new and as yet undeveloped. If Christians are to share in that process of development they will need to . . . be prepared to let go of ideas from the past that are no longer adequate (in Ball and others 1992: 4).

Other contemporary theologians agree: 'What is needed now', says Ruth Page, in *God and the Web of Creation*, 'is not another skirmish on the green fringes of belief but a rethinking of fundamental doctrine' (Page 1996).

Second is the challenge to our comfortable life-styles. Although the probable consequences of the environmental crisis are well publicised in the western world, and the statistics get more alarming every year, few people are willing to face the unpleasant fact that, sometime quite soon, it will no longer be possible to carry on with our lives on the assumption that the future will be a more or less logical extrapolation of the present. Profound life-style changes, especially in Northern* countries (the places where such changes will have the greatest ecological impact), are highly unpopular (McFague 1993: 3), and so are studiously avoided. The necessary task, adds McFague (*ibid*, 17), is to get people to see that it is not enough merely to change our life-styles; we must change what we value, and that will be not only more difficult, but virtually unprecedented.

Third is the perceived (but illusory) conflict between the demands of the poor and of nature. Left-leaning activists tend to be suspicious that attention to environmental concerns will divert scarce resources away from the more immediate issues of poverty and hunger (Hallman 1994), or that environmentalism is a luxury of affluent Northern societies:

> The exploitation of nature and the exploitation of other human beings are inseparable; they reflect a common set of cultural values and a common framework of economic and political institutions . . . Despite the significant legislation it has produced, the conservation movement has been limited in its long-run effects because it has tended to think of nature and man apart from this social context. Its victories are at best stop-gaps in the face of the population explosion and burgeoning industrial pollution. The movement has often been supported by relatively privileged groups, people who could afford hunting and fishing or vacations in National Parks. Conservationists have usually assumed that no fundamental changes in our society are needed. Like John Muir in the last century, they have often scorned the city and instead have urged escape from society and its problems into the beauty of the wilderness (Barbour 1972: 159-60).

There is resentment in the South that the Northern nations, having ruined their own environments, now want to hamper development in the South so as to protect their own life-styles. So people still ask whether money spent on environmental protection would be better diverted to social programmes (Daly and Cobb 1990: 377), or, picking up Barbour's point, they see conservation as a hobby for the wealthy, since only those who had money could afford to worry about dolphins (Granberg-Michaelson 1992: 9). But McFague's description of nature as 'the new poor' (McFague 1993: 166) neatly underlines the link between environmental and social concerns. Christians do not have to make a choice between caring for the poor among human societies and valuing the natural world as the handiwork (however interpreted) of God; both duties are done together when we challenge the ruling economic paradigm which threatens both.

2.1 Ecumenical background

The first and basic requirement for the health of the *oikoumene*, the whole inhabited earth, is that it be habitable (Rasmussen 1996: 91). 'Habitat' or 'household' is the core meaning of all three 'eco' words— ecology, economy, and ecumenism, and all three meanings meet in the wider implications of church-based environmental activism. So it is entirely appropriate that the first glimmerings of interest in ecological matters as a legitimate concern of the church can be dated to an address given by Joseph Sittler to an ecumenical organisation, the World Council of Churches (WCC) in New Delhi in 1961 (Williamson 1992: 93). He pointed out that, ever since Augustine, Western Christendom has been unable to connect the realm of grace with the realm of nature. The universally accepted Greek idea of a dualistic split between the spiritual and the temporal necessarily meant that redemption has had to be visualised as an escape from the cosmos of natural and historical fact.

However, this attitude is isolated between, and inconsistent with, both the Biblical concepts that preceded it and the modern, secular European ideas that have followed it. For the Hebrews, creation was to be understood as the scene of God's sovereign activity, not a prison of evil to be escaped (Simkins 1994). And to modern ears the Augustinian formulation is untrue to the organic character of Biblical language, and unintelligible in the present state of human knowledge and experience. Rather, in modern times we can again insist that the natural world is not simply 'the stage on which the larger drama of history is played, but has a key role in that drama itself' (Granberg- Michaelson 1994: 98).

WCC was founded in 1947, with a largely Protestant background and leadership (Gerle 1995). The attitude to environmental matters of the relevant WCC sub-unit, Church and Society, until New Delhi and also for the next twenty years after it, remained one of low-key, pragmatic instrumentalism. A programme emphasising the Christian vision of a Just, Participatory and Sustainable Society (JPSS) was established in 1975, in response to a warning from Charles Birch to the Nairobi Assembly that, in ecological terms, the world is a *Titanic* pursuing its collision course with the iceberg (Williamson 1992: 93). Although the concept of sustainability is rooted in the Biblical idea of order in creation (Rasmussen 1996: 161), JPSS did not include any idea of protecting creation for its own sake.

That all changed at the Vancouver Assembly in 1983, after Jürgen Moltmann pointed out that the long-established interest of the churches in peace and justice for humans was not enough by itself: there can be no peace and justice for us unless we also protect the natural world. The delegates therefore resolved to call for a new global conciliar process to be established, to covenant for Justice, Peace and the Integrity of Creation (JPIC). Their specific rejection of the previous concept of sustainability in favour of the somewhat woollier and unfamiliar one of 'the integrity of creation* was 'greeted with dismay' by WCC at the time (Gosling 1992: 7). But by then the churches in Germany and parts of Scandinavia and Canada had become heavily involved in environmental issues, and the Pacific churches had long been suffering the effects of nuclear testing. Between them they shocked the Vancouver delegates into a re-evaluation of the entire ecumenical agenda. Hence, although secular concern about nature conservation is by no means new, and research on it is considerably more advanced outside the churches than within them, Vancouver made the unique contribution of declaring that justice, peace and the integrity of creation are to be seen as inescapably linked.

The response to the Vancouver initiative by local churches was delayed while WCC made continued efforts (until late 1987) to get active participation from the Roman Catholic Church.[1] The negotiations failed, largely because the RCC and the WCC had different ideas of what was meant by a conciliar process. The RCC thought of it as a 'top-down' one, exclusively run by and for the church hierarchy*, whereas the WCC insisted on a 'bottom-up' approach including many lay people (Niles 1992). Considering the gravity of the issues at stake, McDonagh (himself a Roman Catholic) calls this a lame excuse (McDonagh 1994: 106).

When JPIC eventually got into gear, the reaction of churches around the world varied according to the current local concerns of each region. For example, the primary concern in the Pacific was the question of anti-nuclear protests; in Africa and Latin America, the unbearable burden of foreign debt; and in southern Africa, apartheid (Gerle 1995: 55). At the first Conference of European Churches, held at Basel in 1989, ancient antagonisms were fading, but only because more urgent problems were rising, to do with the new ecumenical environment. The meeting brought together almost seven hundred delegates from one hundred and twenty member churches of the Conference of European Churches, and the twenty-five Bishops' Conferences of the Council of European Bishops' Conferences. It was the first occasion on which representatives of the Roman and Orthodox churches had officially met since the fifteenth century, and the first ever involving thousands of ordinary people of all the modern European faiths—Roman Catholic, Orthodox and Protestant—who joined the delegates in worship. Hans Küng (1990) called it 'a model contribution', and he listed (67-69) with approval the unusually frank self-criticisms of the participating churches.

Regional variation in dominant concerns is important and valid, because local communities tend to feel threatened by the lack of recognition of local realities implied by too much emphasis on a global perspective (Gerle, 1995). Anyway, the local variation did not obscure the general agreement, at these and at the many other meetings held to

---

1. The Roman Catholic Church was not involved in WCC's earlier assemblies: for example, it did not co-sponsor the important World Congress at Seoul in 1990, and took up only 20 of the 50 observer seats allocated to it there, although it has since done many parallel studies of its own (McDonagh 1994: 106).

discuss the concept of JPIC (listed by Gosling 1992: 15), that the addition of the third term to JPIC, concerning the integrity of creation, should be strongly endorsed. The message was further confirmed by several influential books published during this period, such as *God in Creation* (Moltmann 1985), *Imaging God: Dominion as Stewardship* (Hall 1986) and *Liberating Life* (Birch and others 1990).

The culmination of all this effort was the definitive JPIC World Convocation held in Seoul in March 1990. The key issues at that meeting were the Southern* debt crisis, militarisation, the atmosphere, and racism. The Final Document, reprinted as an appendix to Niles (1992:164-90), lists a series of ten affirmations covering the most important issues of justice, peace and world ecology. The affirmations directly relevant to this book were:

> 7. The creation is beloved of God. We have a responsibility to care for creation, to respect the rights of future generations and to conserve and work for the integrity of creation.
> 8. The earth is the Lord's. Human use of land and waters should not destroy the life-giving power of the earth.

The affirmations were followed by four covenants, by which Christians were called upon to work, among other things:

> 3. For preserving the gift of the earth's atmosphere to nurture and sustain the world's life; for building a culture that can live in harmony with creation's integrity; for combating the causes of destructive changes to the atmosphere which threaten to disrupt the earth's climate and create widespread suffering. The churches can develop new theological perspectives concerning creation and the place of humanity in it, and join the global, local and personal efforts to safeguard the integrity of creation.

The final document from Seoul, strongly influenced by the many representatives attending from the South, describes a worldview grim even in 1990 and getting worse every year, but which is still not high in the consciousness or political agendas of most Northern governments:

> We have entered a new period of history in which humanity has acquired the capacity to destroy itself. Developments in the areas of economics, politics and technology cannot continue on their present course. More and more people are realizing that a radically new orientation is required if catastrophe is to be avoided. Movements of resistance are taking shape in many parts of the world. Such movements are also growing in the churches (Gerle 1995: 63).

Berry (1993b) outlines the history of environmental concern in Britain, and religious responses to it. In general Britons are well aware of problems ahead but not united by the 'sense of approaching catastrophe' which initiated the processes of JPIC. Likewise, the Conference of Churches in Aotearoa New Zealand was quick to point out that the 1990 Seoul meeting coincided with the one hundred and fiftieth anniversary of the Treaty of Waitangi, and their booklet on JPIC strongly urged local discussion of the links between JPIC and the Treaty (Anon 1990a). But its long-term impact on local congregations was microscopic.

Meanwhile, a parallel process of consultation and exhortation was going on within the Anglican Church—which is a member of WCC, but is also large enough to conduct its own enquiries into matters raised by WCC meetings that could be especially important to the world-wide Anglican communion. The most authoritative Anglican assemblies are the Lambeth Conferences, called every ten years by the Archbishop of Canterbury and attended by bishops representing every province in the Anglican world. The Anglican Consultative Council (ACC) was established in 1969 to provide for more frequent and broader-scale discussions between successive Lambeth Conferences.

ACC is an international assembly of clergy, bishops and lay people from throughout the Anglican Communion, and it usually meets every three years. The 1988 Lambeth Conference discussed environmental concerns; and its report, plus the events in Basle and Seoul, were discussed by ACC at its 8th meeting in Wales in 1990. In the final report of that meeting, entitled *Mission in a Broken World* (Anon 1990b: 101-103), the ACC sought to

> bring up to date the definition of mission which has been developing within ACC, and to relate that to the current phase of human history . . . A consistent view of mission repeated by ACC, the Lambeth Conference [and others] . . . defines mission in a fourfold way:
>
> The mission of the church is:
> (a) to proclaim the good news of the Kingdom;
> (b) to teach, baptise and nurture new believers;
> (c) to respond to human need by loving service;
> (d) to seek to transform the unjust structures of society.
>
> We now feel that our understanding of the ecological crisis, and indeed of the threats to the unity of all creation, mean that we have to add a fifth affirmation:
>
> (e) to strive to safeguard the integrity of creation and sustain and renew the life of the earth.

The fifth of these mission statements urges the Anglican Church to be, or become, involved in the world-wide effort by all thinking people, of any faith or none, to find workable ways to alleviate the global environmental crisis. Just like the Seoul declarations, however, the Statement is an incompatible mixture of contemporary scientific and religious environmental concern set against a Biblical background that had no such concern. Public exhortations based on either, taken at face value, are unlikely to succeed, especially if addressed to secular audiences. Does that mean that secular initiatives might do better?

The 1992 UNCED Earth Summit at Rio was an attempt to make real progress on the environmental debate by organising a global consultation at the very highest level (Palmer 1995). It was a huge and chaotic event, divided between an official meeting of Heads of State[2], a Global Forum of over 1,400 non-government organisations (NGOs), and an ecumenical meeting organised by WCC, all on different sites and all with different agendas. Many mainstream churches prepared statements for it, including the Anglican Communion (Berry 1993a: 263-4). Its consequences have included important new resources for conservation provided by national Governments which committed themselves to participating in various forms of remedial action under Agenda 21 (the development of action plans for the 21st century). If there was any common theme there, it was the open rejection of the assumption, seldom questioned in UN circles before, that all the diverse societies of the world could be put on a single track, along which the 'under-developed' countries merely represented an early stage in a desirable and inevitable progress towards the 'developed', western way of life.

> Historically the Earth Summit will come to mark to time when the world realized that development as traditionally understood had failed (Granberg-Michaelson 1992: 1).

The Southern nations insisted that the net flow of resources to the North, due to massive debt repayments and unfavourable terms of trade, would have to be reversed before there could be any hope of initiating programmes for environmental sustainability in the South, and few had any illusions that the unrestrained free market would help. Against this, the Northern countries, especially the US, worked hard to downplay the criticism of their life styles. One of New Zealand's representatives there, Sir Geoffrey Palmer, describes the US performance as 'abysmal' (Palmer 1995), and it was the cause of much

---

2. Some 170 nations were represented by their leaders, but not New Zealand. The official commemoration medal presented to Heads of State was given instead to the Maori Queen (R. Laurenson, pers.comm).

anger as it removed all target dates from the climate convention and refused absolutely to sign the biodiversity convention (Wilkinson 1993).

In one respect, the North prevailed: President Bush simply declared that the standard of living of US citizens was not up for negotiation (Rasmussen 1996:133). However the South, and many Northerners with them, did manage to link poverty reduction with environmental protection and get them put on the global agenda for everybody, even though the basic policy decisions of business and finance were out of reach altogether. At the same time, during the two weeks of the meeting the global population increased by >3,000,000 (to 5,467,000,000) and the total area of productive arable land decreased by 100,000 ha (Wilkinson 1993). Effective co-operation to deal with these problems was sabotaged by international tensions—which were even worse at Earth Summit II in 1997.

The UNCED meeting achieved much less than it might have done: but it did produce Agenda 21, the Rio Declaration and two significant, legally binding conventions on biodiversity and climate change. A third document, intended to become a binding convention on forests, was reduced to a Declaration because the conference could not reach agreement on the wording (Palmer 1995). Many local authorities have developed Agenda 21 programmes, while New Zealand is a national signatory to both Conventions, as well as to the separate 1987 Montreal Protocol (Hay 1996).

The Seventh Assembly of WCC in Canberra in 1991 urged the continuance of the work of JPIC (Niles 1992); but Rio showed that the massive participation of NGOs in the debate is now so advanced that a church-based ecumenical movement no longer needs to play a leading role in supporting and stimulating such groups (Granberg-Michaelson 1992: 47). In 1994 the WCC reorganised its administration structure, and JPIC was incorporated into a new programme renamed 'Theology of Life' (Chial 1996). 'What has yet to emerge', comments Chial, 'is a theology that significantly inspires change' (*ibid*, 58). This is a lot to ask: the current JPC web page (http://www.wcc-coe.org/wcc/what/jpc/ecology.html) has information on peace and justice programmes, on globalisation, biotechnology and climate change, but hardly mentions the theology of creation at all.

Fortunately other forms of international consultation on the relationships between religion and the environment have been developing alongside those of the Christian churches (Baker 1996). In 1986 the World Wildlife Fund sponsored a meeting of religious leaders in Assisi, to discuss how each of their communities of faith could contribute towards stimulating environmental awareness and promote conservation within their own traditions (Berry 1993b). The result was renewed interest in studies of how the various sacred writings teach respect for the earth; the initiation of thousands of conservation projects and environmental education programmes around the world; and the creation of the Network on Conservation and Religion (Anon 1995). In 1990 the Joint Appeal by Religion and Science for the Environment was established, followed in 1993 by the National Religious Partnership for Environment, based in New York. In 1995, a Summit on Religions and Conservation in London, again sponsored by WWF along with the Pilkington Foundation and a Japanese humanitarian foundation, brought together religious leaders representing nine of the world's major faiths to review progress since the 1985 meeting. In February 1998 the Archbishop of Canterbury and the President of the World Bank hosted a 'Dialogue on World Faiths and Development' at Lambeth Palace (www.worldbank.org).

All of this talking and meeting is a step in the right direction, although (as was obvious at Rio) it is unclear to what extent the exhortations of those concerned about global survival could influence the national policies of those concerned to maintain or improve their own standard of living. The most prominent example of this problem current at the time of writing was the refusal of President George W Bush to ratify the Kyoto Protocol in March 2001, which had been signed by his predecessor Bill Clinton and about 80 other government leaders, because it was 'fatally flawed' and bad for the US economy. Perhaps it is no coincidence that it was George W's father who had derailed several important agreements at Rio, on much the same grounds.

2.2 Why earnest exhortations are never enough

Despite the strong convictions conveyed by scientists, UN organisations, religious and green activists and other authors for at least the last thirty years, the cold hard fact remains that practically nothing has changed. There has been no very obvious response by people, congregations, corporations or governments, and the global crisis continues on its way as before. We urgently need to know, why have these perfectly serious efforts by official and otherwise respected organisations apparently had so little effect? As usual with such a large question, there are many answers. But my suggestion is that activists and moral agencies of all sorts tend to underestimate the importance of recognising the cultural and historical environment in which they operate. In particular, according to Primavesi (1991), church authorities tend to perceive their teachings as pure deposits of truth handed down though generations of social, political and cultural vacuum. To any student of the human animal, this is nonsense.

Managing the environmental crisis is partly a moral problem, partly a political and economic one, and partly a scientific one. There are, then, three overlapping spheres of primary knowledge that are relevant to it (Fig 2). No appeals for action that ignore these other considerations, especially those based on purely religious arguments, are likely to be either intellectually respectable or effective in practice.

The human environment is only a small part of the whole creation of God, but since (so far as we know) we are the only creatures capable of caring for any of it, our concern should extend to all of it. Hence, the idea of environmental management, usually considered a secular discipline with a variety of sub-disciplines, is merely a subset of the wider task of caring for creation. However, even the more limited aim of environmental management is difficult enough, for two main reasons. First, the natural world is far more complicated and unpredictable than our theoretical models can handle, which means that errors in management decisions are far more common than we like to admit (Budiansky 1995; Caughley and Gunn 1996). Second, the human environment is a common resource for all humanity, and management of it is not the business of one or a few individuals but a matter for collective action by many people whose interests usually do not coincide.

For both reasons, management of the environment, like that of any common property, is not a simple task. Organising fair and just

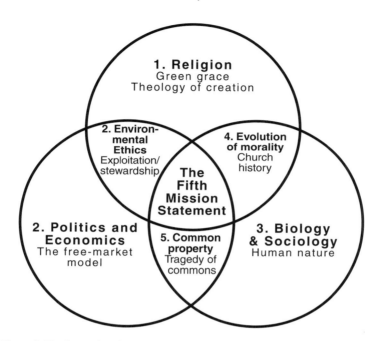

Figure 2. The three cultural contexts in which a church-based concept such as the Anglican Fifth Mission Statement, or any other form of religious environmental activism, has to operate.

collective action among a population of independent egoists is a problem that has exercised secular philosophers, legislators and lawyers for centuries. Their disciplines have histories of their own, reflecting their independent development particularly within the liberal intellectual atmosphere of western civilisation over the last two centuries. But since the UNCED conference at Rio it has become clear that Northern style economic growth and progress are not normal and cannot be continued indefinitely. The environmental crisis is not merely a problem of applied ecology; it has wider and deeper dimensions, which are a lot more frightening and so are explored less often:

A century ago Thoreau could truly say, 'There are a thousand hacking at the branches of evil to one who is striking at the root'. This is not as true now as it was in

Thoreau's day, but whether we are more successful in eradicating evil is questionable. We have trouble recognising a major root when we see it (Hardin and Baden 1977: 5).

These interactions also have a time dimension. It is useful to distinguish between the short term, localised environmental dilemmas that face all Northern societies, including that of New Zealand, and the longer term global crisis that threatens humanity as a whole.

The short-term dilemma is that the development and implementation of environmental policies within a western democracy is largely, though not entirely, a matter of common property management. The associated problems are not new: there is a well established branch of conventional economic theory specialising in it, and a vast literature on the culture and management of communal resources (eg Hardin and Baden 1977; Keohane and Ostrom 1995; McCay and Acheson 1987; Stevenson 1991). On the local and short-term scales on which we as individuals all live, common property will probably always have to be managed according to some sort of updated version of this theory, so we must learn to develop and monitor the best means of doing that. On the other hand, the longer term crisis cannot be addressed within the framework of conventional macro-economic theory, because it is that theory itself which is at issue. The associated problems are completely new, and indeed not officially recognised by many economists, so the literature is so far confined to a few prophetic and unpopular voices such as Ehrenfeld (1981) and Daly and Cobb (1990). They, in their turn, are often half-drowned in the 'brownlash' from the vested interests they criticise (Ehrlich and Ehrlich 1996). We need to remember both the short-term and the long-term perspectives, together and separately, and their implications for the churches' hope to make a contribution to the environmental debate in the twenty-first century. The theology of economics is a huge subject in itself: here I concentrate on the *biological* reasons why the challenge to humanity, to learn to live within its limits, is at the same time so obviously necessary and also so impossibly difficult.

The main problem is that the shift to sustainability is not just a matter of calculating the resources we can each use; it is also a challenge to society and to individuals to overcome the materialistic attitudes and values which are responsible for the over-consumption and environmental damage in the first place. We in the North read authoritative reports such as that of Bruntland (1987), documenting the growth of material consumption and of population that inevitably generates ever-greater crowding, pollution and degradation of resources. We watch news film of poverty, famine and conflict in distant places in the South, and realise that western society cannot totally disclaim responsibility. We understand more or less that the free-market model, which works by systematic privatisation of profits to individuals or small groups, coupled with (wherever possible) socialisation of costs to the most vulnerable human and natural systems, inevitably means that all our technological advances have been achieved by the progressive sacrifice of two of humanity's most treasured traditional assets: a supportive local community and a healthy, productive natural environment. The loss and degradation of the natural environment is a clear threat to long-term human survival (Paul Ekins, in Daly and Cobb 1990:vii). We applaud the critics of the dominant, free-market model such as Galbraith (1992; 1996), Chomsky (1998) and Daly and Cobb (1990), and relish quotes from them such as the oft-repeated one from Kenneth Boulding–himself an economist : 'Anyone who believes that exponential growth can go on forever in a finite world is either a madman or an economist' (Anon 1994: 55).

Unfortunately, we are less willing to admit that the root of the problem of the total human environmental impact lies with us, the people of the industrialised North, simply because we consume so much more per head. The fifty-seven million Northerners born in the 1990s—a mere five per cent of the total global population growth—will be responsible for more environmental degradation during their lifetimes than the nine hundred and eleven million extra people born in the poor Southern countries during the same period (Athanasiou 1996: 37). Another way of putting it is to point out that the 2.9 million people in Chicago (or the 3.6 million in New Zealand) consume about as much per year as do the ninety-seven million in Bangladesh (Rasmussen 1996: 39). The clear conclusion is, therefore, that responsibility for working towards sustainability lies very heavy on Northern peoples. We are the only ones in a position to do it, and we could do so without risking our own survival, by reducing our often excessive consumption. As the old saying has it, we should live more simply, so that others may simply live. But will it happen? I doubt it.

The prospects of the people of any Northern democracy achieving such a heroic level of self-sacrifice in their own living standards in order to assist the poor of the South seem remote. Intellectually, we may be able to agree that it is in our own best interests to do so, but in practice, the chances are slight to zero. Even changes that are urgently needed to improve conditions in our own environments are resisted. Why should this be?

Certainly, all arguments of the form 'if everyone co-operated to recycle bottles/organise car-pools/avoid wasting water/ate less meat/refused to buy eggs from battery hens' (or whatever other personal sacrifices are seen as needed to do good in the world) are bound to fail. The contradictions between private benefit and public good are all around us. For example, Athanasiou (1996: 42) points out that, even though European cities are clotting with traffic, no amount of analysis of the benefits of public transport negate the facts that people want cars, and the freedom and social status associated with them, just as they want beef, TV and the good life generally. The threat to the economics of running a private vehicle which arose during the Gulf War, and which might have made public transport or alternative fuels more attractive, merely demonstrated that Northerners would 'rather fight than switch' (*ibid*, 24). The kind of radical overhaul of economic policy that Daly and Cobb, and even more so Hawken (1993), are calling for would require a social consensus and an act of collective will and determination totally unprecedented in peace-time history. Many thoughtful people have concluded it is impossible.

The main reason is that the changes required strike at the core of human behaviour. The transformation of personal values required to accept a less materialistic life style is a matter of education and spirituality, not economics (Granberg-Michaelson 1992: 49). But all the while economic advantage as defined under the prevailing paradigm remains the conscious governing motive defining rational behaviour in our society, and ignorance of our evolutionary background prevents analysis of our unconscious attitudes, no such transformation is likely.

Statistics bear out this rather gloomy prognosis. In a 1993 study of consumer attitudes in Germany, quoted in the report *Towards a Sustainable Europe*, G Scherhorn identified four distinct sets of lifestyles among the German population, which the editors of the report considered would probably be representative of trends in other European countries as well (Friends of the Earth Europe 1995: 202). Only one group, comprising twenty per cent of the population, were considered to be both aware of environmental and social problems and also prepared to take responsibility for doing something about them. These people tended to have a high income and education, and to be what Scherman called 'post-materialists', ie they had developed independent personalities and a comprehensive set of intrinsic values, so did not depend on material assets to provide the security and self-esteem that are the usual pre-conditions for the ability to appreciate the natural environment.

The other three groups were all labelled 'pro-materialists' and classed as unlikely to take responsibility for either social or environmental problems, because they were entirely materialistic in their outlook (twenty-five per cent), or believed they could achieve a high level of material welfare without paying the price in environmental damage (thirty per cent), or were totally apathetic (twenty-five per cent). The fact that these pro-materialists comprised about eighty per cent of the population means that, as the editors well realised (*ibid*, 203), an immediate change in life-style and values is highly improbable— even in Germany, long considered to be one of the most environmentally aware countries in Europe.

To understand the reasons for the apparently irresponsible attitudes documented by Scherman, we must look more closely, first at why the management of common-pool resources (including most environmental goods) is particularly difficult (p.45), and then at what evolutionary biology can tell us about the nature and sociality of humans (p.74).

3. Environmental Management and the Tragedy of Unmanaged Commons

Almost all the important aspects of the natural environment which are at risk in New Zealand are resources common to all New Zealanders—rivers,[1] lakes, groundwater, air, endangered species, biodiversity, national and forest parks, leasehold tussock grassland, offshore islands, inshore fish and so on. So also are many of the human creations of the social environment that must be managed wisely, such as access to taxpayer-funded civic institutes, art galleries, health care and education. On the wider scale, there are other forms of resources common to humanity in general, such as ocean fisheries, whales, the ozone layer, Antarctica and so on. In order to appreciate why the well-meant admonitions of green activists are so often ignored, it is necessary to consider the special problems of managing common property.

3.1 The problems of common property management

There is a continuum of types of property, ranging from private property (as in the exclusive possession of one's own house) to no property (as in the open oceans and outer space). In between there is common property, where the rights of access, possession or exploitation are shared by individuals in common with others. These rights may allow unlimited exploitation for those in a specified group, as in access to the internet by those who have paid their ISP fees, or they may stipulate limits or quotas on each user, as in commercial fisheries: but in either case there is definitely no open access.

The resources that are commonly owned may themselves be of two kinds, which represent the extreme ends of a spectrum of competition for access (Keohane and Ostrom 1995). Public goods are those that yield

---

1. At least, they have been regarded as common property until recently. Current arguments over, for example, the claim of the Tainui tribe to ownership of the Waikato River, are stimulating new discussions on which resources should belong to all New Zealanders and which should be returned to the ownership of the historically dispossessed tribes.

infinite benefits, in the sense that additional use by A does not diminish the amount of benefit remaining for B. No matter how many people photograph Mitre Peak and breathe the oxygen in the fresh air of Milford Sound, or observe the local weather forecasts, navigation marks and lighthouses, the amount of each available to the next person is the same, so there is no competition for them. At the other extreme are common-pool resources which offer subtractive or finite benefits—if A uses more, less remains for others. Competition for common-pool resources is real, and if extensive, they become affected by depletion, congestion or degradation—the net results of any form of use pushed to beyond sustainable limits. The Milford road, the car park overlooking Mitre Peak, the beach at the edge of the Sound, and space for tourist boats at the jetty are therefore all common-pool resources, even though the view of the Peak remains a public good.

The difference between public goods and common-pool resources is significant, because some resources change from one type to the other over time if their normal superabundance becomes threatened by human activities (for example, fresh water in the Auckland area was a public good until the local human population passed a certain density, but the cumulative destructive effect of over-use, waste disposal, fertiliser runoff etc has since made it a common-pool resource), and some vary with locality (oxygen becomes a common-pool resource in a closed environment such as a submarine or aircraft cabin). Public goods generally cause rather little environmental concern unless they make the transition to common-pool resources. The term 'use' includes both taking resources out of an ecosystem and putting waste products into it.

The problem of managing common-pool resources can be stated as follows. Consider a group of people who are placed in a situation where they could mutually benefit if all adopted a rule of restrained use of a common-pool resource, such as petrol during the 'oil shocks' of the 1970s. Game theory (p.85) helps to explain why, even though it is in all their interests to co-operate to conserve the supply, they will not unless everyone is made to do the same by some recognised agreement or external authority. All individuals acting severally have an incentive to maximise their private use of the resource and ignore the associated social costs, not because they themselves are especially greedy, but for fear that others will benefit from their restraint. If there are reasons for ruling out price as a factor to regulate use, then various sorts of rationing must be imposed—one solution introduced in New Zealand in the late 1970s was the 'car-less days' scheme. Common-pool resources with open access, like roads, or fixed price, like basic foods, or no price, like biodiversity, cannot be controlled by market mechanisms, and neither does the market provide any incentive to protect public goods, for example to keep the oceans clean or protect the ozone layer. The inevitable result is that the aggregate use exceeds the renewal rate of the resource (Wade 1987).

Garrett Hardin (1968) long ago suggested that management of common resources is difficult under certain conditions, which unfortunately are quite usual. The argument can be illustrated from the example which gave us the word 'commons': the medieval idea of a village pasture. In that system, a number of people held joint ownership over a common resource, such as common pasture sufficient for fifty villagers. They all had equal rights of access or use, and all had the right to graze their own cows on the common. Hardin's argument assumed that all the villagers made their own independent decisions, and that in theory there was nothing to stop Bloggs grazing two cows. His profit would be one whole cow, whereas his loss would be 1/50th of the damage to the pasture. If everyone made the same calculation, the total use of the resource would be bound to exceed the supply, and the pasture would inevitably become overgrazed. The result is what Hardin called *The Tragedy of the Commons* (a tragedy in the sense of being both bad and inevitable). It happens wherever people are free to privatise benefits while socialising costs; economists know it as the 'free-rider problem'.

As many authors have pointed out, and as Hardin himself later acknowledged, this parable is a caricature of the way that real common grazing lands were managed in medieval England, and still are, in equivalent contemporary situations in, for example, alpine Switzerland (Stevenson 1991) and elsewhere (Ecologist 1993). In practice, there are always socially-enforced restrictions on this process, which even in medieval times had been developed by local agreements over centuries. In a small village everyone knew what everyone else was doing, and

Bloggs' self-interest would be immediately detected and prevented. But when those social restrictions against access to a common-pool resource are removed, and especially if personal dealings can be done in private, there is nothing to prevent the rational decisions made by the individual users each acting in their own best interests, and out of sight of all the others, from eventually damaging everyone's interests. It happens all the time to other kinds of public resources held in common by much larger communities where social cohesion is weak or non-existent, especially in modern times, because Bloggs' contemporary equivalents can conduct their calculations under a cloak of anonymity.

Therefore, the system Hardin describes is actually industrial-system logic in rural dress, with nature regarded as free goods whose use is regulated by the market. It is not real commons-logic, where shared resources are regulated by community decision (Rasmussen 1996: 338). Where access is open, Hardin's description of the chain of events is valid, but, as he suggests himself, his parable should be renamed 'The Tragedy of the Unmanaged Commons' (Hardin 1994). The difference between an open-access system and a true commons, in which real authority rests with a real community, is all the clearer for the contrast. The key conclusion from a more general statement of Hardin's parable, very relevant to this book, is that physics and biology ultimately set limits to the options available for moral and political life (Elliott 1997).

I have collected together the following examples of the tragedy of unmanaged commons, taken largely from modern New Zealand, and presented them in the form of three 'laws' which I have labelled with three well-known names. The names are attached not because these individuals formulated these 'laws' themselves, but because their actions help us to understand the general principles summarised by the 'laws'.

3.1.1 First form: Muldoon's Law

In management of a common resource, strategies that are individually rational can be collectively disastrous

Modern economic law practically deifies the rights of individuals to pursue their own best interests by any legal means, but at the same time it frequently fails to mitigate the effects of the result on the common-pool resources at risk. Those effects are likely to be bad for the population in general if the private, rational short-term strategies of individuals have collective consequences that are destructive over the long term. The most obvious examples concern the apparently unlimited common resources of the past, which were taken to be public goods until they were destroyed or irreparably damaged by individuals claiming their rights within the law of the time: for example, ocean fish, seals and whales, natural tussock grasslands and forests.

Muldoon's Law is irrelevant to the management of genuine public goods, but swiftly punishes communities that mistakenly or selfishly ignore signs of limitation showing that public goods are making the transition to a common-pool resource. The early pioneer settlers of Canterbury could discharge smoke and human waste into their environment without a second thought, but their descendants in Christchurch do not have the same freedom now that clean air and water can no longer be regarded as unlimited resources.

Muldoon's Law is especially likely to apply if the common resource is believed (truthfully, or through self-deception) by individuals to be an effectively unlimited public good, and who like the pioneers see no need for personal restraint, even though the supply of the resource is in fact limiting to the group as a whole. The classic example is that of the man for whom I named the process, the New Zealand conservative politician Robert Muldoon. His 1975 election campaign was based on a promise to implement a generous and unearned national superannuation scheme, funded by an unstated tax-based contract of mutual support between generations (the present workforce pays the retirement pensions of the present elderly on the assumption that they will get their own in turn when the future workforce does the same). The voters of the time (as now) treated the national tax revenue as a public good, so they enthusiastically supported a measure that proposed to give everyone over sixty a generous share of it whether they needed it or not. Muldoon achieved his intended result (he was elected), but at the price of contributing—along with many other problems—to unintended, long-term damage to the New Zealand economy (Kelsey 1997: 24) and considerable inequity between generations (Thomson 1991).

These days it is very clear that no fisheries, forest resources or taxpayer funding is ever unlimited, and new regulations quite properly restrict individual access to them even though they are still common property. But we seem not to have learned the lesson; other forms of common resources are still subject to Muldoon's Law if they meet the basic conditions for its operation, which are (1) open access, especially (2) by very large groups in which individuals may act privately, and especially (3) if the long-term effect is actually very damaging while the people concerned believe, or convince themselves, it is not. Even normally honest people may be willing to cheat on a large business, especially if it is making a substantial profit, on the grounds that 'they can afford it'. The key point is that open access is appropriate only for a genuine public good offering non-subtractive benefits, like the view of Mitre Peak—and even then, the access route to and amenities around the main view point are not public goods.[2]

3.1.2. Second form: Berk's Law

The threat of damage to or depletion of an uncontrolled common resource increases its value and stimulates competition among free individuals to harvest it all the faster, regardless of the future

Commercial exploitation of uncontrolled natural resources (here, 'uncontrolled' includes also 'insufficiently controlled') is almost always disastrous. In the earliest pre-colonial period of New Zealand history, huge populations of breeding fur seals on the subantarctic islands were wiped out by commercial sealers between 1792 and about 1820 (King 1990: 255). In modern times, such practices can be halted by legislation, although usually not until after prolonged battles. For example, few important fisheries are uncontrolled, but the sizes of the remaining stocks are often overestimated and the quotas have in the past been too generous. Fishermen, faced with declining catches, only try harder. Why should this be so, since it must be obvious to everyone that unrestrained exploitation has the effect of killing the goose that lays golden eggs?

---

2. I do not intend to make unhelpful and unfair criticisms of individuals for contributing to Muldoon's Law; rather, to question the economic model that guarantees their rights to do so.

First, many resources increase in value as they become rarer, and increased value is usually sufficient in itself to guarantee increased competition for access. Second, if the resource is not properly controlled, for example if it is in international territory like the air or the ocean, open access and scramble competition ensure that even if individual A holds back from exploiting the resource, others will not, so the stock continues to decline. A's restraint conveys no benefits, neither private nor public, so Berk's Law proceeds even if everyone can see what is happening. Third, the economic rule of present-value maximisation dictates that even a genuine golden-egg-laying goose should be killed if the interest earned by a private owner on the capital gained from selling the carcase exceeds the future value of the eggs (Daly and Cobb 1990: 156)–and even if the community is thereby impoverished of the precious birds. Fourth, self-deception is a powerful ally of idiocy and greed in these situations.

It is widely assumed that modern commercial exploitation is more damaging than the hunting practised in traditional societies, for several reasons. First, and most significantly, there is no effective limit to the financial incentive that drives commercial hunting, which is a quite different process to subsistence hunting. Second, commercial hunters are not dependent on a particular resource for survival in the same sense as traditional hunters were—when the resource runs out they do not starve, because they can turn to other work even if it means moving or retraining. Third, the control of the group over the behaviour of individual members that was characteristic of the small, close-knit tribe, and which is crucial for real commons management, has broken down in modern society.

But in fact the supposed contrast between modern commercial and traditional hunting may not be real, since the only forms of traditional exploitation that have survived to the present day are the ones which have proved to be sustainable over the long term. Traditional hunting of resources that were extremely sensitive to exploitation, such as that of the early Polynesian settlers on the moa in New Zealand, and of many other colonists on other island bird faunas (Quammen 1996), was in its time just as drastic in its effect as has been that of modern whaling and cod-fishing. On Easter Island, the people who cut down the last living

51

tree must have known that it was the very last one, since the island is so small, yet powerful social forces made sure they still cut it (Bahn and Flenley 1992). At first, when the supply of moa and other birds seemed unlimited, there was nothing to prevent massive overkill; later, when Berk's Law came into effect, there was nothing to restrain hunters driven by desperate hunger.

In *The Origins of Virtue*, Ridley (1996) argues against the romantic idea that indigenous peoples are necessarily any better conservationists than Europeans: if they have done less damage in the past, that is because their technology was less destructive. He quotes (223-4) Nicanor Gonzalez, the leader of a South American indigenous people's movement: 'We aren't nature lovers . . . at no time have indigenous groups included the concepts of conservation and ecology in their traditional vocabulary'. For example, the Kayapo Indians, given control over a twenty thousand square mile reserve in Brazil, were soon enthusiastically selling concessions to gold miners and loggers. In other words, when other pressures are or were strong enough, and the resource fragile enough, Berk's Law can work just as well in any society. Tim Flannery's analysis of the problem is aptly entitled *The Future Eaters* (Flannery 1994).

If the 'resource' is valuable to some members of the community but regarded as a pest by others, unrestrained commercial hunting can serve the interests both of the hunters personally and also of the common good. For example, for about fifteen years after the discovery, in about 1970, that it was possible to shoot deer, tahr, goats and chamois from helicopters (Caughley 1983), the intensive exploitation of these animals over huge areas of inaccessible high country in New Zealand had a devastating effect on their populations and a net beneficial effect on the native vegetation. That was not the intention of the hunters, but the effect was positive anyway. Far more often, the resource is valuable to everyone, and unrestrained exploitation by a few deprives everyone. I named this idea after the common description of an idiotic profligate as a 'berk'.

So why do people not co-operate to stop Berk's Law progressing, either in the past to prevent the extermination of an open-access resource such as the fur seals, or in the present to manage the current environmental crisis? Because, as game theory so clearly demonstrates (p.165) it would be irrational for A to hold back from using the resource if there is nothing to prevent B, C and all the rest from taking it if A does not—everyone else gains at A's expense, and the resource continues to decline faster and faster. When people convince themselves that the resource is bound to decline anyway, they become all the more determined to be sure of getting their share before it is too late.

The classic examples of this process are the nineteenth century collectors of rare birds in New Zealand, who knew very well that many species were declining into extinction, and that their activities accelerated the process, and yet continued to collect them in huge numbers:

> Since the species were bound to die out, then obviously specimens should be taken while they were still available. Buller liked collecting—the thrill of acquisition, of having the lovely, rare, feathered specimens. As their rarity increased, their value in the eyes of fanciers like Buller and Rothschild soared (Galbreath 1989: 207-8).

In modern environmental decison-making, the influence of self-deception is both widespread and deeply puzzling to philosophers. The wider questions it raises about the origins of morality and of self-deception in human affairs are discussed later (p.109).

3.1.3 Third form: Bolger's Law

Individuals will tend to resist restriction of private access to common resources, even to protect the long term interests of the community

Recent New Zealand governments have inherited the huge national debt created partly by the operation of Muldoon's Law during the decade from 1975. The Labour Government elected in 1984 introduced a massive structural adjustment programme, driven by free-market economic ideology in its purest form. It included a surtax on the generous payments of Muldoon's pension scheme, such that everyone still received it as the law required, but those who had more than a minimal level of other income paid back part or all of their national superannuation in tax. This measure met predictable outrage from voters, who thought of the national tax revenue as an effectively unlimited public good and resisted the surtax as an intolerable infringement of their personal right of access to it. During the general election campaign of 1990, Jim Bolger promised to repeal the detested surtax, but once safely in office as the next conservative Prime Minister, he merely replaced it with an even more draconian income test (Kelsey 1997: 287). People have remembered and resented Bolger's betrayal with amazing ferocity, even though most can also see that Muldoon's original scheme was unsustainable in the long run, and that therefore it was in everyone's interests that something be done about it. Rather similar arguments have been heard since the originally very generous payments of the New Zealand Accident Compensation Commission have had to be curtailed.

Another equally graphic illustration of Bolger's Law in operation was played out in Rotorua in 1987. The city stands above a large but limited reservoir of underground geothermal fluid, which supplies the famous geysers and other geothermal features of the region. The very important local tourist industry is quite directly influenced by the health of the geysers, especially Pohutu—the largest geyser in the Whakarewarewa field, itself the last of New Zealand's five known active geyser fields still remaining in its native state. In the late 1970s, concern was rising about the decline in Pohutu's activity, due to excessive withdrawal of the geothermal fluid through the many private bores in the city. Eventually, in 1986 the Government took emergency action, and implemented a progressive programme of bore closures. In 1987-88 the programme accelerated to include compulsory closure of all private bores within a 1.5 km radius of Pohutu, plus all Government wells in Rotorua. It was obviously in the interests of everyone in the local community to protect the tourist trade, but nevertheless the owners of the private bores were outraged. There were large public protests, resistance (to the extent of attempting physically to prevent access to the privately-owned bore-heads), and legal challenges to the Government. Subsequent monitoring of the aquifer water level and pressure, and the activity of Pohutu, showed significant improvements attributed to the closures, vindicating a Government decision considered by many at the time to be 'harsh, draconian and un-necessary' (Grant-Taylor and O'Shaughnessy 1992).

Examples of Bolger's Law could easily be multiplied, because the restriction of private access to public resources is becoming more and more necessary as human pressure on the environment increases. In both New Zealand and the western states of USA, large areas of publicly owned grassland have been leased to private farmers for pastoral grazing since the last century. When animal pests (rabbits in New Zealand, and coyotes in USA) threatened to make such farming uneconomic, many farmers demanded removal of the pests at public expense, resisting any suggestion that the public benefit might be better served by a policy of rethinking the whole idea of uneconomic farming on public lands. Small local communities often protest when their long-established access to the free natural cleansing powers of the sea is ended, and replaced by compulsory payments (via higher rates) for sewage treatment. Perhaps the ultimate example is the one-child-per-family policy in China, in which the public resource at stake is living space and a sustainable life for the existing population. In the long term that can be protected only by restricting the number of new individuals entering into it, and that in turn can be achieved only by drastic state interference into the private lives and decisions of individuals. One does not have to experience the stresses of life in China to agree that the policy, or something like it, is necessary; but at the same time one does not need to be a Chinese to sympathise with the individuals who resist it.

A more dangerous form of Bolger's Law operates when the individuals resisting restriction of access to the resource are also the ones who make the rules. They are then in a position to do a great deal of damage. In highly stratified societies controlled by a relatively small elite, such as in eastern Europe under the communists, one or a few powerful people can manipulate the government institutions that keep social arrangements intact even when the cost is obvious and there is progressive environmental deterioration (Rasmussen 1996: 42). Ironically, among the worst examples of this comes from the other side of the political divide. In the early days of the Reagan administration, the Secretary for the Interior James Watt refused to restrict commercial access to natural resources in US because he believed the Lord was liable to return at any moment, so it would be pointless to implement environmental protection that could hamper US industry in the meantime (Swadling 1989).

In effect, maldistribution of power and social inequality permit powerful forces to benefit without hindrance at the expense of both the people and the resource until it is too late. CS Lewis made the same point, decades ago, when he commented that what we call human power over nature is actually the power exercised by some people over others, using nature as a tool. The Marxist version of the same dictum is even more succinct: 'power over nature becomes power over people' (Barbour 1997: 144). Once again, the theme emerges: justice and equality are essential pre-requisites for any future protection of nature—down to, and including, the level of local authority policies about litter (p.163).

3.2 Theories of collective action

Economists refer to the tragedy of unmanaged commons as a 'market failure', reflecting their customary assumption that market control of public assets is or should be the norm (Hartley 1997). Market failures are usually caused by uncontrolled 'externalities'—profits or losses not accounted for in the market model—and corrected by vesting of property rights, or by institutionalising arrangements for 'internalising the externalities'. These seldom work as intended—and anyway the free-market philosophy has limited application to many forms of public assets. So it is worth emphasising that there are, broadly, three alternative strategies under which the usage of a common-pool resource may be regulated. All involve the co-ordination of the individual decisions of the users by various degrees of agreement or coercion, but the circumstances favouring them vary. The moral challenge is to find ways of making the coercion as painless, as humane and as unobtrusive as possible (Elliott 1997).

3.2.1 Privatisation

In a private environment under individual or family ownership, the owner has freedom of control over the resources it contains. The effect of privatisation is to internalise the benefits of good management, which therefore adds motivation for improvement policies including, for example, the reduction of overuse. Rational management is in the user's (ie, owner's) own best interests, so most property owners are careful about keeping up with maintenance and preventing damage. Most of the benefits of access to private resources are internalised to the owner, although they sometimes also have a side benefit for the public; for example, any passer-by can enjoy the sight of a well-kept garden or farm supporting beautiful flowers and wildlife. Conversely, the costs of bad management are borne directly by the user. The net result is that benefit and cost are both significant when it comes to making management decisions, and ideally they should be more or less balanced—provided the owner is committed to long-term dependency on the resource, and cannot abandon it without penalty for another one when it is damaged.

When the owner depends on the resource, the rights, responsibilities and survival value of ownership become linked; therefore private resources are generally well managed, which ideally should encourage good conservation of resources over the long term. The eighteenth-century enclosures of common land, and the huge advances in agricultural productivity that followed them, were driven by the insistence (not necessarily explicit) of progressive landowners that the benefits of land development should be internalised, ie should benefit only the owner and not the labourers. The disadvantage of private ownership is that it leads to gross social inequalities of benefit, and eventually to envy and injustice. After the enclosures, the productivity of English farmland soared—but the improvements benefited only the few who had secure private ownership of good farmland. A far greater number of labourers were dispossessed of all access even to communally owned land, and in time fell into great poverty (Ecologist 1993: 25-6).

Conversely, in open-access public environments such as a national highway system, all individual members of the group have at least theoretical freedom of access to the resources owned by the group. Rational management is certainly in the group interest, but it is not the responsibility of any individual user. The benefits of access go to *individual users,* while the costs of bad management are *divided between all users.* The net result is that the rights and responsibilities of ownership are divorced; and since the benefit to any one individual from exploitation is much higher than their share of the cost of it, long-term environmental damage becomes a real danger. The main advantage of public ownership is that it allows social equality of benefit (or loss) in the short term.

If private resources (such as land, houses, and businesses) are usually better managed than public ones, because of the natural self-interest of the owners, the question arises: could self-interest help to conserve public resources too? Adam Smith, in *The Wealth of Nations* (1776) promoted that view in his famous declaration that an individual who 'intends only his own gain' will be led by, as it were, *'an invisible hand* to promote . . . the public interest' (p.98). Smith's metaphor was memorable, and although it was mentioned only in passing, the message it so neatly encapsulated suited powerful landowners and politicians, especially those to the right of centre, so it was and still is widely believed. Hartley (1997) applies the modern version of the same logic to the management of conservation lands in New Zealand. But is it true?

In practice it all depends on the scale of the operation and the number of people involved. Muldoon's Law (p.48) illustrates why unrestricted self-interest may manage private resources well, but it manages communal resources badly, unless they are owned and used by a very small social group all of whose members are very well informed about what all the others are doing. How small a group can effectively shortcut Muldoon's Law?

Hardin (1993: 266) quotes the collective management of land by the Hutterites, who live in religious colonies of between sixty and one hundred and fifty people. As colonies grow, more and more members become 'drones', defined as those who attempted to maximise the only personal luxury allowed in that society, leisure time. In a colony of below one hundred and fifty members, close social scrutiny makes shirking difficult, but in larger ones it becomes easier. Hutterite colonies are therefore deliberately split up when they exceed that number.

Privatisation also tends to rule out certain forms of activity that might be in the common interest but whose benefits cannot be internalised. The classic example is the lighthouse (Ridley 1996: 102): who will pay for a lighthouse when the light is free to all users? The same argument applies, pro rata, where the services needed are not free but still uneconomic, such as those of rural post offices and buses. Rasmussen (1996: 326) asks: 'How efficient and realistic is it to continue with . . . "dumb design", that is, design that never asks what the health of ecosystems and human communities require and that results in horrendous waste and injustice?' Under the prevailing free-market model that dominates world finances at the moment, the logic of privatisation answers: If it makes a profit, it's efficient and realistic.

Regardless of all criticisms, the privatisation model will not go away. After all that has been said above about the failings of the free market, any suggestion of extending it in order to advance the interests of conservation must seem completely illogical. But Hartley (1997) argues the case at great length, and specifically in the New Zealand context. He concedes that, in some cases at least,

> it is undeniable that externalities or a positive demand for public goods can make it extremely difficult, if not impossible, for markets to attain the *maximum conceivable* benefit from the available resources [his italics]. However, political and central planning approaches to resource allocation have their own limitations and failures (viii).

The problem, says Hartley, is not failure of markets, but *insufficient markets*. Control of common resources by a management agency on behalf of others via parliamentary legislation (the usual pattern, eg of the New Zealand national parks by the Department of Conservation) raises questions of management accountability and incentives. It also encourages people to find ways of transferring to others the costs of meeting their aims—for example, by exaggerating the demand for, and the benefits of, politically provided goods and services. In a clear reference to conservation activists, he specifically singles out for criticism the claim that the environment is sacred, or otherwise incommensurate with commercial activities, which he sees as a ploy used to pre-empt alternative resource uses. Because conservation is a costly activity, users of conservation assets should be required to pay the full cost of their claims on valuable resources, he says. For example, the present policy of providing free access to conservation areas reduces the cost of operating a commercial eco-tourism venture and discourages development of full-cost private resources for that purpose. 'If DoC were a commercial organisation, its current pricing policies would be illegal' (x). Methods of controlling demand other than by prices, eg by refusing to provide easy access to remote sites of interest to potential tourists, are inefficient.

By contrast, Hartley says, all these problems could be resolved by intelligent and creative use of property rights, which has great potential to aid the conservation of habitats and species and encourage sustainable resource use. The example he gives is a familiar one:

> The disappearance of the moa in New Zealand
> provides another example of the disastrous effects of
> lack of private ownership of valued resources (xii).

Hartley's analysis was published with the backing of the New Zealand Business Roundtable. Perhaps they discouraged him from considering any system of thought other than market economics; whatever the reason, the outcome is that Hartley made many errors of judgement about conservation problems, including the following.

1. Hartley's questioning of the accountability and incentives of public-service conservation managers betrays total unfamiliarity with the motives, integrity and attitudes of the people who actually carry out those responsibilities.
2. The first national park in New Zealand (Tongariro) was gifted to the nation in 1887 by its Maori owners, led by Te Heu Heu Tukino, precisely *because* the sanctity of that environment was to them a

deeply held conviction, not a ploy in a political argument. If the gift had not been made then, for reasons which are totally incomprehensible in Hartley's view, no conservation agency could afford to buy the mountain at today's market prices.

3. The ideal of free access to national parks goes back to the very foundations of the earliest national parks in US and Australia in the 1870s, whereas the free-market model dates only from the Bretton Woods conference of 1944.

4. In some countries, the idea of charging park user fees has been reluctantly accepted in areas of very high potential use, not for economic reasons but mainly because unrestricted access would cause damage to the natural values being protected—user charges can easily cost more to collect than they bring in.

5. The argument that the extinction of the moa was due to 'lack of private ownership' is simply absurd. The moa of any given area could be hunted only by the tribes that claimed communal ownership of that land, and they disappeared because, like all large endemic birds, they bred very slowly and were extremely vulnerable to any sudden increase in mortality (King 1984).

The then Minister of Conservation, Nick Smith, rejected Hartley's recipe for managing New Zealand's conservation lands as 'based on greed and exploitation' (*Forest and Bird*, February 1998, 4). But within months came the next round of arguments in the same vein: a critique of the Resource Management Act, commissioned by the Minister for the Environment and written, in provocative style, by an independent consultant, Owen McShane. The call for public submissions on McShane's report received seven hundred and fifty responses (Anon 1998), and the argument continues. The pressures exerted on public assets by economic interests, disguised as instruments of the common good, may have been moderated somewhat by the 1999 election result, but it remains to be seen what changes the new Labour-Alliance government is able to implement.

3.2.2 Regulation

Not all common-pool resources can or should be privatised. The next alternative, therefore, especially where increasing scarcity of resources is bringing Berk's Law into action, is regulation. Hardin (1993) is one of the best-known advocates of hard-headed state interference in even the most sensitive personal decisions, such as family planning. As he points out, one of the most pressing of the world's problems is over-population. Some authorities maintain that it is the only real problem, and all others are consequences of it. Others disagree, on the grounds that it is not the actual numbers of people that matters but their lifestyles, particularly in the highly consumptive Northern world (p.41). As the American attitude to the Rio and Kyoto proposals clearly demonstrated, people are not likely to take kindly to regulations suggested by other cultures affecting their family lives or standards of living. Yet, says Hardin, such regulations will soon be seen as inevitable—preferably in the form of 'mutual coercion, mutually agreed upon'.

In one of his best-known metaphors, Hardin likens individual nation-states to lifeboats, some of which are too overcrowded to stay afloat. But, he asks, if you are in a lifeboat that can carry fifty people, and fifty more are in the water, what do you do? If you take in all fifty, then one hundred will drown when fifty might be saved. If you take in ten, you have to decide which ten, and what to say to the other forty. If you take in none, you will be safe but you still have to face yourself later. If you and forty-nine other altruists offer your places to others, then all altruists will drown and the total surviving remains the same—all of them more selfish than the average. (This is a particularly distressing conclusion to Christians schooled in the virtues of self-offering love). In a strictly enclosed system, the dilemma is insuperable. Hence Hardin concludes

> To couple the concept of freedom to breed with the belief that everyone born has an equal right to the commons is to lock the world into a tragic course of action (Hardin 1968).

Hardin quotes with enthusiasm Kenneth Boulding's idea that in a future world, girls will be issued with tradeable child-bearing certificates when they reach fifteen, and those who produce more children than they have certificates for will be compulsorily sterilised (Hardin 1993: 273). The chances of any such scheme being introduced into any Northern democracy in the forseeable future seem negligible. It would be not only politically impossible, at least until the effects of the environmental crisis become a lot more obvious than they are yet, but also counter-productive. Experiments show that people's willingness to exercise self-restraint is affected more by their communication with each other - that is, by whether or not they live in a functioning community–than by the prospect of punishment. 'Covenants without swords work; swords without covenants do not . . . so much for Hardin's plea for coercion' says Ridley (1996: 240).

Wherever the main or only form of communication possible is personal, the optimal size for a community is up to about 150 (Hardin and Baden 1977). This rule has applied throughout human history until the last century or so. However, now, in the age of electronic communications, much larger communities are possible that could have the same effect, and they might favour greater degrees of mutual regulation in future. The advent of the Internet might improve the prospects of widespread consultation on and mutual agreement about matters of public good within very large communities. If not, or for resources that are too widespread for management by any form of personal involvement, Oye and Maxwell (1995) offer a moderately optimistic view of the prospects of developing rational environmental policies by imposing 'systems of regulation and compensation that bring about convergence of narrow self-interest and the common good'. Of course, where the interests of the individual and the group coincide, regulation need be only minimal: the more they conflict, the tighter the regulations have to be to protect the common good. Oye and Maxwell therefore distinguish two different kinds of regulation.

In what they call 'Stiglerian' situations, the desired convergence of self-interest and common good is a by-product of regulations constraining competition, limiting entry by new competitors and encouraging monopolies. Because the benefits are conferred on a few and the costs diffused across the many, the few who are regulated benefit from regulation: they lobby for it, and the system, once established, is stable. For example, the Montreal Protocol that banned the use of CFCs created a market for substitute chemicals in which DuPont and ICI had a market advantage, so, after opposing regulation on principle for as long as they could (Athanasiou 1996: 64), these companies strongly supported the regulations which benefited both their own commercial interests and the common good. Similarly, the phasing out of leaded petrol created a larger market for the higher-profit unleaded petrol; the banning of DDT, a cheap and easily-made chemical produced independently in many Southern countries, promoted a move to more specialised substitute chemicals which are more difficult to produce and so favoured the interests of larger (Northern) companies; and restrictions on new housing developments and enterprises such as salmon farms favour the owners of existing houses and installations.

On the global scale, there are few effective international authorities capable of controlling access to or use of global commons in order to force a Stiglerian solution. The action of the UN/US during the 1991 Gulf War, portrayed as a means of protecting the sovereignty of Saudi Arabia and Kuwait, was very obviously driven by the self-interest of the Northern powers, which could not tolerate the idea of Saddam Hussein gaining control of the oil wealth of the friendly Arab states. All the same, the congruence of interests of the UN/US and the Saudis/Kuwaitis achieved the desired effect, perceived to be in the cause of the global common good, of terminating Hussein's territorial adventure into Kuwait. Less dramatic forms of UN regulation can also be effective, but experience confirms the comment of the World Commission on Environment and Development, that 'hammering out an international consensus [on managing the commons] . . . is a huge task requiring time and patience' (Bruntland 1987: 286).

In 'Olsonian' situations, the benefits of regulation are diffused across the many while the costs are borne by a few. No natural convergence of interests is possible, so it must be created artificially. Those regulated will often protest, and the only way to achieve stability is to arrange for compensation from the many to the few. Regulations that are seen to be unfair can be unstable, although if the compensation is substantial it can be decisive in achieving a solution. In one case,

> the Hokkaido Electric Company paid 300 million yen to one of four small fishing co-operatives opposed to the construction of a nuclear power plant. Not surprisingly, payments of this magnitude diminished local opposition to the plant (Oye and Maxwell 1995: 215).

Loggers and farmers can be relied upon to oppose restrictions on cutting rights imposed to protect endangered species. In one well-known New Zealand case, at Pureora in 1978, payments for logging contracts lost or broken by government directive (but only as a result of determined protests by conservationist groups) were reported to have cost NZ$7 million—a substantial sum at that date (Wilson 1982; Wright 1980). In Britain, under the Wildlife and Countryside Act 1981, taxpayers compensate owners of private resources for not implementing developments on their own land that would benefit the owners but be detrimental to the public good (Moore 1987). The idea is laudable, but, as any cynic might predict and as Oye and Maxwell observe, the supply of compensation may generate its own demand, and encourage deliberate deceptions by local landowners who may not have intended to do the disputed developments at all but threaten them as a cheap means of extracting compensation.

Olsonian compensation does not seem likely on the global scale, at least not as yet, but a form of it has certainly been suggested, based on tradable rights. For example, one suggestion is that every country should have only a certain number of rights to emit pollutants or greenhouse gases such as carbon dioxide, and therefore the rich North, which pours disproportionately more carbon dioxide per head into the global atmosphere, should compensate the poor South for taking up the South's allocations as well as their own (Richards 1991). Such allocations, based on already-recognised population data and territorial authority, would give the South an unfamiliar and welcome bargaining advantage based on justice rather than on aid, reverse the present flow of wealth from South to North, and go far to alleviate the South's unbearable burden of debt. Needless to say, Richards' idea that, 'for the first time in history the disinherited may have a grip on the powerful' would be fiercely resisted by the North, and—as amply demonstrated by recent events—it has so far proved impracticable to impose on any wealthy democracy. On the local level, regulations such as 'user-pays', which attach replanting obligations to forest cutting rights and clean-up obligations to mining permits, are small steps in the right direction.

Unfortunately, systems of regulation imposed by the state are not only expensive to administer, but they have been known to misfire, especially if they take control of a resource out of the hands of a local community and put it in the hands of an inefficient government bureaucracy. An example recounted by Ridley (1996: 236) is the action taken a few years ago by many African countries to nationalise their wildlife and game reserves, on the assumption that nationalisation would be the best way to protect this common-property resource from poachers. But the actual effect was that local farmers suffered damage from government-owned elephants and buffalo, without having any balancing incentive to look after the animals as a source of meat or tourist revenue. The decline of African elephants, rhinos and other animals outside the parks was, says Ridley, in part a tragedy of unmanaged commons caused by nationalisation. But the situation can be rapidly reversed whenever title to wildlife is re-privatised to local communities and hunters have to bid for the right to shoot game. Then the villagers change their attitudes to the now-valuable game on their land. The acreage of private land in Zimbabwe devoted to wildlife increased from 17,000 to 30,000 km^2 after the government granted title over wildlife to landowners. Likewise, Nepalese irrigation systems run by the state or by aid agencies are less efficient and less equitable than those run by the villagers themselves (Ridley 1996: 237).

Such cases illustrate, says Ridley (1996: 262-3), that where authority replaces reciprocity, the sense of community fades, so heavy government makes people more selfish, not less. Therefore, systems of regulation, however tempting, are often not the best answer, for a reason he summarises in a few words: 'Ecological [and economic] virtue must be created from the bottom up, not the top down' (*ibid*, 246).

3.2.3 *Collective action*

The drawbacks of both privatisation and imposition of regulation by an outside authority invite closer inspection of the third means of managing a common property, by genuine collective action. The possibility of long-term success in such action is often dismissed, largely because of the huge and continuing influence of Hardin's parable, despite criticism (Vink and Kassier 1987 quote one reviewer as saying that 'it would be difficult to locate another passage of comparable length and fame containing as many errors'!), and of similar theoretical models such as the Prisoner's Dilemma (Stevenson 1991; Wade 1987). Yet empirical observers of societies where common-pool resources have been successfully managed for centuries, such as Switzerland, India and South Africa, point out that theoretical models simply do not explain reality. Wade therefore set out to ask what conditions favoured, or denied, successful collective action.

At one extreme, he says, we should not expect to find any system of restrained access organised by the users themselves when:

1. there are too many independent users;
2. the boundaries of the common resource are unclear;
3. when the users live in large groups scattered over a wide area;
4. when undiscovered rule-breaking is easy.

Such conditions assisted the virtual extermination of seals and whales in New Zealand waters during the late eighteenth and early nineteenth centuries (King 1990). The same can be said with reference to the management of social resources at the nearer end of the historical scale, the abuse of taxpayer funds as summarised by Muldoon's Law described above. 'In these circumstances', says Wade, 'degradation of the commons can confidently be expected as demand increases, and privatisation or state regulation may be the only options'. But both of these have their disadvantages. So are there any other alternatives?

Wade (1987) lists a number of definable circumstances under which self-organisation by user-groups to manage a common resource can be successful. I have illustrated Wade's list with New Zealand examples.

1. The resource is relatively small and located within definable boundaries, but the technology to privatise it (eg by fencing) is expensive. In New Zealand the Himalayan tahr (a goat-like mountain animal) and sambar (a large Asian deer) are introduced game animals much admired by hunters, and both occupy reasonably well-defined but unfenced ranges on public lands. The hunters who value them lobbied against excessive control of tahr by the Department of Conservation in the late 1980s (Hughey and Parkes 1996), and organised a voluntary ban of sambar hunting when their populations fell too low in 1982 (extended indefinitely when it expired in 1987) (King 1990).

2. The users live close to the resource, depend on it for survival and understand about sustainable yields. The user group is small, clearly defined, democratic, with high mutual solidarity and a system for noticing and chastising rule-breakers, and has a leadership which benefits from good management of the common resource. In the Classic period of Maori culture in New Zealand, each of the many distinct tribes held their marine and forest resources in common ownership. The complex regulations of *tapu* controlling the harvest of shellfish, finfish, birds and timber (Best 1942: 132-5) were regarded as a matter of life and death to the tribe, taken so literally that early European visitors such as Marion Dufresne who ignored them were attacked in rational self-defence (Duyker 1994).

3. The state either does not exist as a separate institution, as in New Zealand before 1840, or it exists but does not interfere with legal forms of collective action, hitherto always locally-based. In the modern world, for the first time, awareness of environmental problems is international, and *global* collective action is now possible. One of the most successful forms of it ever seen has been the world-wide boycott on the trading and use of animal furs. The rationale of it was more to do with the prevention of cruelty to animals than the management of a common-pool resource, but it remains an impressive example of influential collective action driven by an intense emotional commitment and coordinated by world-wide communication. The consequences have been serious for New Zealand, where the Australian brush-tail possum, an introduced fur-bearing mammal, is a serious pest. The devaluation of furs on the world

market has made trapping uneconomic, thereby removing one formerly important means of controlling possums. The international ban on whaling, though less successful, is more directly a reflection of world-wide public concern over a vulnerable common resource.

Where the above three conditions are met, collective action often works. Some of the immense variation in effective forms of local authority over local resources, built up over the course of history in traditional societies, are discussed by Hardin (1977) and Ecologist (1993). The latter book, *Whose Common Future?* is a protest against the 'paternalistic outsiders' and the webs of international power that run most development programmes stimulated by the Bruntland Report—together with a passionate appeal for the restoration of genuine commons management.

Because the conditions favouring collective action are so variable, there is no guarantee that it will necessarily work in a given situation, any more than privatisation or state regulation necessarily works. The dismal frequency of degraded natural commons, such as despoiled forests, overexploited groundwater and depleted fisheries—not to mention social commons such as chronically overcrowded roads and vandalised public conveniences—shows only too clearly that collective action cannot always be presumed to be effective. It all depends on the circumstances. The scheme proposed by Wade (1987) emphasises only that the probability of descending into such anarchy or destruction is neither as strong nor as general as the theoretical models imply. His conclusion is that, where the circumstances may favour collective action to manage a resource, we should not be too hasty to reject it as an option, especially since it may be cheaper to implement and more stable in the long term than either privatisation or external regulation.

3.3 Conclusion

Management of common resources, including many aspects of environmental management, involves a difficult balancing of individual and public interests. There are at least three possible ways of doing this, but no simple universal rules. Which is the best option depends on the conditions, and most particularly on the size and mutual solidarity of the local community. Application of game theory can often help understand different people's reactions to a given scheme (p.163). The only thing that all three management options have in common is that none of them will work without education, monitoring and social or legal sanctions to back up rules or collective agreements. This common factor is the main reason why the well-intentioned declarations emanating from Seoul and Rio have had no general effect. Green activists need to understand that before adding to the already massive world stock of useless exhortations.

The ultimate challenge for the development of an environmentally responsible life style in the modern world is to foster a widespread congruence between personal and group interest, which is the basis of community responsibility or 'public spirit'. Achieving that congruence has always been one of the social functions of religion (Burhoe 1979). When a communal cultural story is widely accepted and regularly reinforced through rituals such as worship services, the group is seen to be more important than the individual. Part of the reason why community responsibility seems to be waning in the contemporary Northern world is that the traditional stories underlying Christian society have lost their influence on individual ideals and behaviour, and the more basic forces of human nature are freer than they were. Why should that be so? Perhaps a deeper understanding of human nature might help to explain how people and, especially, politicians in western society no longer value public virtue as much as they once did, and what the consequences of this loss might be.

4. Human Nature

4.1 If there is a crisis, why are people ignoring it?

It is obvious that everyone ultimately depends on the resources that creation provides, and therefore everyone has an interest in ensuring that creation, and especially the part of it that constitutes the human environment, remains healthy. So why do the earnest exhortations of theologians and green activists so often fall upon deaf ears? As Brennan (1993: 18,8) asked, 'If we really desire to pass on the earth in good shape to our children, then why do we not act on this desire? . . . [why don't] decision-makers and policy analysts . . . take account of our individual and corporate tendencies to tell ourselves comforting stories that help us live with self-deception and weakness of will'? These are among those simple questions to which there are no simple answers, but a lot of clues can be found in recent advances in evolutionary biology.

First, well-intended exhortations must be realistic. There must be some irreducible minimum of damage to creation, simply because of the sheer size of the human population that it has to support. Humans now dominate all earth's ecosystems: they monopolise somewhere between thirty to fifty per cent of the land area of the earth[1], about sixty-six per cent of recognised marine fisheries, more than half of all accessible rainwater, and in places more than that of irreplaceable fossil groundwater (Vitousek and others 1997). No-one, least of all any Christian, is seriously suggesting a deliberate campaign to kill off surplus people; those who already exist are accepted to have the right to continue to exist. Therefore the problem of caring for creation is confined to (a) mitigating the burden of supporting the present human population, and (b) minimising the consequences of that burden for future generations. There is plenty of room for improvement in both (and for suggestions that there should be a third aim, (c) to progressively reduce the size and impact of the global population), but

---

1. 'Monopoly' here includes both transforming or degrading a resource by human enterprise, and excluding other species from using it.

all such ideas meet some formidable obstacles.

These obstacles are partly in the nature of things. For example, the energy equations of plant productivity are much the same in all ecosystems, but energy moves faster through some plant communities than through others. Much of the primary productivity of a forest is locked up in long-lived and inedible plant material, such as trees, whereas the primary productivity of a field is short-lived, rapidly replaced and easily converted into bread or meat. Therefore, it is simply impossible to support as many people on the products of a forest as on the products of open grassland. That is the reason so much forest has been cleared, from historical times onwards, and why there is a practical limit to the proportion of forested land that any populated country can afford to reserve—even though many endangered species evolved in, and can live only in, large areas of forest.

But such purely technological matters are minor considerations besides the really daunting obstacles to conservation imposed by the nature of humans. Even given the very large global human population, caring for creation would be technically possible, even though difficult, if only everyone could agree on what needs to be done *and co-operate to do it*. The fatal dilemmas that dog the best intentions of WCC and the international conservation movement alike arise from the fact that our *individual* attitudes and behaviour often sabotage the *collective* decisions we need to make to implement the wise management of our common environment on a sufficient scale. That alone makes caring for creation a *moral* issue, one that transcends the objective discussion of how ecological damage to our environment arises. Sustainable development is *not* 'business as usual with green bits' (Lawton 1994); taking the concept seriously forces people to realise that they have to make agonisingly difficult *choices*. Ecological morality is a matter of weighing up the consequences and significance of one's own actions, which can be labelled as good or evil in terms of their effects on other people, countries, biota or ecosystems, and then making conscious, personal decisions:

> You either decide that most human beings in the
> developing world can never own cars, or you have to

imagine a future in the developed world with many
fewer cars—in fact a developed world with rather less
of most things . . . so what will you give up? Who, in
western democracies will vote for less, not more?
(Lawton, 1994).

In turn again, those choices presuppose a clear idea of what
constitutes morality, what distinguishes good and evil, and where these
concepts came from. For the Christian biologist, that means
understanding evolutionary biology and the relationship between the
evolutionary and the cultural stories of humankind.

The main outlines of the current theory of evolutionary biology are
standard knowledge, but recent advances are frequently
misunderstood, and when it comes to applying them to human life there
is plenty of room for different interpretations.

The most important of these is the debate about the levels at which
natural selection works within human social life–to what extent are
contemporary humans subject to its workings? Are we simply the
'robots' of Richard Dawkins' provocative metaphor, the product of
unconscious genetic mechanisms over which we have no control—or,
worse still, the unknowing hosts of parasitic memes that manipulate
human behaviour to serve only their own interests (Blackmore 1999)—
or do we have real choices? Can humans consciously deflect the
processes of natural and cultural selection in order to develop and
practice public virtues? It is easy to give reasons why we need to combat
the damaging assumptions about the self-interested behaviour of
individual *Homo economicus* which are the basis of the economic system
that has governed western society for the last two hundred years (Daly
and Cobb 1990). Unfortunately, biology has some sobering things to say
about the extent to which we can hope for help from human nature on
improving human attitudes to the global environmental crisis.

A second, equally important debate is about reductionism*. The
main problem of biology is to explain how so many complex structures
have been built up in the world from apparently simple beginnings.
Dennett (1995) distinguishes two kinds of theory, illustrated by two
mechanical metaphors. A 'crane' is an openly understandable,

scientifically testable theory, such as evolution by natural selection. A 'skyhook' is a non-testable hypothesis appealing to outside help, either from aliens or from God. In Dennett's view, critics of natural selection (mostly non-scientists) are unable to give up hope in skyhooks because they cannot quite believe that cranes can do the job unaided. Among scientists, he distinguishes between 'greedy reductionists' and 'good reductionists'. The greedy ones try to do without *both* skyhooks and cranes, ignoring intermediary mechanisms including natural selection and descending immediately to physics. The good ones are opposed only to skyhooks, and their work is the basis of all modern biology.

To understand the arguments I present, it is essential to understand natural selection as it operates in the natural world. In Appendix 1, I offer a summary of the main facts and processes of natural selection, using where possible New Zealand examples seldom found in textbooks. Readers familiar with the contemporary theory might want to check on which of several possible interpretations I use; otherwise, they can safely read on.

4.2 Understanding the roots of human nature

Anthropology has amply confirmed that human social behaviour has evolved in gradual stages from that of our primate ancestors. Foley (1996) identifies eight key 'events' in this prolonged story, starting with the origins of the earliest sociable anthropoids (the monkey/ape lineage) about thirty-five million years ago through to the development of agriculture and the end of purely genetically based evolution between about thirty thousand and ten thousand years ago. For thinking Christians, this is no longer a contentious issue, although few have had the chance to work out the implications for Christian theology or for rational attitudes to the environmental crisis.

The technical definition of humanity is difficult, since human characteristics appeared slowly and over a succession of descendent species. According to Foley's scheme, which is widely accepted among anthropologists, the hominid lineage developed a social system centred on male kin-bonding between fifteen and five million years ago, an almost human upright gait by four million years ago, stone tools by two million years ago, a rapid enlargement of the brain between three hundred thousand and fifty thousand years ago, and the techniques of cave-painting by thirty thousand years ago. The last migration out of Africa and the full colonisation of the rest of the world by true humans with language and agricultural skills was completed between fifteen thousand and five thousand years ago. Sumerian writing appeared six thousand years ago. The Bible takes up the story from about three to four thousand years ago.

The concept that there are no theory-free data is a truism, but is no less true for that. All observations are interpreted through the framework of a particular paradigm that defines what sort of questions may be asked, and what sort of answers might be acceptable (p. 22). It is important to appreciate this when considering any branch of science, but when it comes to the business of understanding ourselves, the limitations of science are at the same time more important to understand and more difficult to escape.

In a science that has for some time been governed by a stable, fruitful and widely accepted paradigm—such as the earth sciences since the establishment of the theory of plate tectonics—there is a broad consensus of agreement among practitioners about the interpretation of data, and few arguments about really fundamental issues. But the scientific study of human nature has emerged only recently, and it has certainly not yet reached such an orderly state, if it ever will.

In order to understand the human problems faced by any form of green politics, we have to understand the biological bases of human morality. The neo-Darwinian perspective on human nature can be extraordinarily illuminating. Many other authors have also discovered new insights from 'Darwinian history'. For example, Colinvaux (1980) used an analogy with the idea of the ecological niche to explain the rise and fall of the classical empires; Crosby (1986) used natural selection to explain the historical success of European colonial expansion; Ridley (1993) used it to explain the pivotal role of sex in the evolution of human intelligence; and Beitzig (1992) used it to examine the sexual propensities of the Roman emperors.

Equally important, we have to understand the differences between the competing paradigms through which human nature may be interpreted. Even within the group of those who agree that evolutionary biology explains the *ultimate mechanisms* that predispose us to behave in certain ways (such as population geneticists, who concentrate on events at the level of the gene), there is no agreement as to how biologists should interact with those from other disciplines who are more interested in the *proximate reasons* why we behave in those ways (such as sociologists and philosophers, who concentrate on events at the level of the individual or the community).

In the following sections, I introduce the ideas of multi-level* selection and of emergent properties as an explanation of the biological origin of morality. Later, I discuss the possibility and the potential consequences of applying the evolutionary perspective to a contemporary understanding of the Christian theology of creation (p.137).

4.2.1 *The primate heritage*

Past a certain stage of development of intelligent self-awareness, rational and sociable animals begin to be conscious of conflicts of interests between one individual and another, and between the interests of any one individual and of the group. Very few of the higher primates are solitary, or have monogamous family units on the human pattern. Most live in sociable groups in which the basic unit is the adult female and her dependent young. Among all the 181 or so species of primates, the exact composition of the group is very variable between species and depends on definable ecological parameters, but in general, the distribution of females follows that of resources; and the distribution of males follows that of females (Wrangham and Peterson 1996).

By comparing various primate species we can identify the conditions that favoured, first, groupings of breeding females, and then, the addition of one or more males. Comparisons with human life must be made cautiously: for example, gibbons are almost always monogamous, and individual gibbons are more faithful to their partners on the average than we are, but the food resources and social environments of humans are very different from those of gibbons. However, it is possible to learn a great deal about ourselves and our ancestors from careful studies of primates, especially our closest relatives, the chimpanzees.

Understanding their implications for human sociality and moral dilemmas is one of the ultimate purposes of the many comparative studies and long-term observations described by primatologists such as de Waal (1989; 1996), Byrne and Whitten (1995), Runciman and others (1996) and Wrangham and Peterson (1996).

The higher primates have greatly advanced capacities for social learning and for reciprocal altruism, and individual members of the group easily recognise each other and remember past favours and abuses, friends and rivals. Chimps have separate hierarchies of rank among males and females, and high social rank is inevitably associated with higher reproductive success. Social hierarchies operate by and reinforce reciprocal altruism. The alpha male is not necessarily the strongest, but the one that is most skilful in forming and manipulating coalitions (de Waal 1982).

Male chimps form coalitions for political purposes: an aspiring male will cultivate powerful friends and use them to support political manoeuvres, or cushion the effects of a social fall, in an astonishingly cynical human way. Not for nothing is the chimp described as possessing pure 'machiavellian intelligence' (Byrne 1995; Whiten and Byrne 1997). Bands of chimps will attack members of other bands, and have been known to completely wipe out rival groups (Wrangham and Peterson 1996). This apparently un-necessary aggression to strangers is inseparable from sociality, which involves both friendly co-operation and regulated rivalry between members of the same group.

Every individual chimp has an investment in its group, and could not survive without it, so the altruism that it will extend to its own companions, which is very definitely not available to outsiders, is hugely important. Individual captive chimps that have had a fight will reconcile afterwards with human-like affection; often a third party will get involved in a clear and apparently conscious attempt to defuse the tensions that inevitably arise in a restricted space (de Waal 1989).

Chimps have also developed not only more varied and context-sensitive personal interactions between the members of one group, but also systems of detailed cultural inheritance between the generations of one group–not instead of genetic evolution by natural selection, but, through cultural selection, in addition to it. Humans still carry the echoes of the basic patterns of primate behaviour which were forged through the millions of years of our pre-human ancestry, and the evolutionary adaptations of our ancestors still strongly influence our attitudes.

4.2.2 The interactions between natural and cultural selection

The processes of natural selection, operating with ruthless logic upon individuals, rule the world of nature. Humans too depend on the natural world for food and shelter, just as animals do, and we still live much of the time by the legacy of our animal ancestors—the rule of 'look after number one'—which can have both good and bad effects (Table 1).

Table 1 Consequences of selection processes in animal and human worlds: mixture of good and bad.

Effect	Natural world	Human world	Scripture
Benefits self and community or other species	Oxygen from plants	Trade	1 Kings 10:29
	Decomposers	Food webs	Lev. 26:3-4
	Pollinators	Pets	Gen 30:15-30
Benefits self and own kind	Reciprocal altruism	Service clubs	Great Commandment
		Adam Smith's "invisible hand"	Mat 22:39
Neutral – benefits self, no cost to others	Epiphytes	Stamp collecting	Meditation
Benefits self but damages own kind	Cancer cells	Tobacco advertising	Achan Jos 7:1
		Tax cheats	Zacheus Lk 18:8
Benefits self but damages community	Dutch Elm Disease	Deforestation	Disobedience to the Covenant Deut. 29: 19

Therefore, natural selection also works in human life. It acts on individuals and slowly, over many generations, since the only definition of success is the greater survival of useful genes compared with others (Appendix 1). We share the same basic biochemistry with virtually all life on earth; the same skeletal plan with all other vertebrates, the same reproductive mechanism with all other placental mammals, and about ninety–eight per cent of our genes with our closest relative, the chimpanzee. The past workings of natural selection explain many aspects of human affairs, from back trouble and flat feet, hereditary diseases and senility, to the earliest stages of the evolution of morality. Cultural selection is a closely similar process that works with memes, mental units of information or ideas (Blackmore 1999). It can act very rapidly, and on groups as well as individuals. Cultural selection can be seen as an advance on natural selection, since unfit ideas can be eliminated without destroying the body that holds them (Table 2).

Table 2 The similarities and differences between natural and cultural selection. Both are essential for understanding human behaviour.

	Natural selection	Cultural selection
Units of selection	**Genes**	**Memes**
Combination with group selection possible	Rare (social insects only)	Common in humans
Sources of variation	Random mutation in DNA Recombination	Sudden inspiration Communication (conversation, publication etc)
Response to need	Very unlikely	Driving force
Mechanism	Interaction chance/ fitness /environmental demands/competition	Interaction chance/ planning/ cultural norms/competition
Consequences	Differential survival of physical characters	Differential survival of ideas
Cost of rejection	Death (personal, or reproductive)	Change ideas
Pace of change	Very slow	Rapid
Extinction	Always permanent	May be temporary

Cultural selection also tends to be less ruthless, because it can ignore the strings attached to all forms of altruism in nature, which rule out any form of animal welfare state. The effects of New Right economics on many aspects of human commercial life, which include the elimination of the unfit (that is, any body, corporate entity or individual trader, that fails to make a profit) are more reminiscent of natural than of cultural selection. Popular opposition to modern economic theory surely includes unconscious resistance to the idea of allowing human society to revert to being governed too obviously by natural selection.

The main difference between immaterial ideas and material genes as units of selection is that ideas are optional, and can be mixed, diluted, acquired, modified or dropped by individuals during their lifetime, whereas genes are given, unitary, specific, inherited by strictly physical processes, and can be acquired only at conception. They cannot naturally be changed or lost during life, and any changes made to an individual, except by new medical technology, cannot be passed on to their children.

But like genes, ideas coexist in different combinations in different individuals, and easily outlive them. Like genes, ideas can mutate into new forms, which then have to compete with existing forms for survival. The sources of variation in ideas can often be, as in genes, a chance observation or a random event. Unlike genes, ideas can also be deliberately developed in answer to a need, and steered in a pre-determined direction. The result of this interweaving of natural and cultural influences on behaviour has been, eventually, the evolution of true, conscious human morality. But because humans are still animals, in the human world natural selection on individuals and cultural selection on groups are mixed, with complex results.

Failure to recognise the important similarities and differences between natural and cultural selection (summarised in Table 2) can lead to some extraordinary conclusions. An extreme example is the work of Ralph Burhoe (1970) who actually identifies God with natural selection, pointing out that both are interested in long-term improvements of types. This may be true, and certainly is an interpretation of religion that suits a scientific culture, but Burhoe fails to notice the differences between the processes employed. Natural selection's method of 'rejecting that which is bad' involves the death or reproductive failure of the individual, which is surely rather a different matter from the Psalmist's prayer that God might 'see if there be any wicked way in me, and lead me in the way everlasting', presumably by personal transformation (a cultural process). Like Richard Dawkins' scientific materialism, Burhoe's evolutionary naturalism is a metaphysical interpretation imposed upon the data, which has to be tested in terms of how well, or badly, it fits in with alternative interpretations. Neither deals adequately with religious experience or historical revelation, so neither has much appeal for Christians (Barbour 1997: 264).

4.3 The theory of gene-culture co-evolution

The model that best allows for the interactions of natural and cultural selection in humans recognises the subtle but important distinction between the levels of selection operating in humans. Natural selection is still there in our unconscious, acting at the level of the individual to choose between rival alleles*. But in addition, we have systems of conscious morality, which have evolved by the different process of cultural selection, and do operate largely for the good of the group. Moral systems have developed as part of our nature, to smooth the constant conflict of interests between members of the community. The seeds of conflict are always there when we do things that favour our own genetic self-interests over the cultural interests of the group, and an intuitive recognition of that conflict clearly lies behind the Augustinian idea of original sin (Campbell 1975). But in close-knit social groups, all individual behaviour is a continual series of compromises that usually benefit the cohesion of the immediate group. Campbell points out that

> for every commandment we may reasonably hypothesize a biological tendency running counter to some social-systematic optimum (*ibid*, 243).

For example, the fact that it served the genetic self-interest of hunter-gatherer males to share their kills (because successful hunters gained social kudos and thereby more sexual partners) did not alter the consequences, which were good for the group.

Cultural selection among humans allows a high degree of group definition, strongly reinforced by the tendency of each member of the group to adopt the behaviour of the majority. This conformist transmission of the learned characters that define group identity acts to increase both within-group solidarity and between-group differences, and so to reduce the probability of defections from the group for personal advantage (Boyd and Richerson 1985: 227). This unlearned propensity for within-group altruism is, of course, the basis of the *esprit de corps* of all cultural units such as schools, colleges and regiments. Human within-group altruism is not independent of biological altruism, but it is at least as closely governed by cultural influences as by genes. At the same time, culturally maintained groups organised into a metapopulation* (eg football teams or college rowing crews, or even more so, national teams at the UN or the Olympic Games) have an unlearned predisposition for between-group rivalry, so they naturally compete. These are the conditions which allow group selection at the cultural level, which in turn has profound consequences for understanding the behaviour of humans.

Gene-culture co-evolutionary theory starts from the assumption that cultural transmission must have preceded the earliest stone tool traditions in *Homo habilis* by a considerable length of time. So humans and their hominid ancestors must have been reliably inheriting two different, well-organised systems of information, genetic and cultural, for at least two million years.

Bowker (1995: 17) points out that there is a spectrum of strong to weak theories of gene-culture interaction. At the genetic (Strong) end, genetic control of culture is all but total and natural selection is supreme. Culture merely serves the genes' struggle for survival, so genes hold culture on a leash. At the cultural (Weak) end, culture is seen as a separate information system with its own evolutionary history, hooked to the genes at various points but not tied to them. Culture has extended, or even slipped off, the leash. Bowker (35) classes early sociobiology, which was at that stage 'regrettably triumphalistic'[2], at the Strong end, and the selfish-gene metaphor, which postulates separate inheritance of genetic and cultural data, at the Weak end. The profound difference between these interpretations explains the otherwise inexplicable fracture of several long-established university departments of anthropology into a cultural branch and a biological branch which, in some places, is severe enough to make communication between them impossible—at least until certain senior staff members move on.

The concept of prolonged interaction between genetic and cultural evolution helps to explain the human story. Boyden (1987) recognises four distinct phases in human history:

> the *primeval hunter-gatherer phase*, lasting for tens of thousands of generations;

---

2. Jeff Schloss, at the June 1998 Templeton conference at Berkeley.

the *early farming phase*, which began independently in the Mediterranean, China and America around 400-500 generations ago;

the *early urban phase*, which began with the first cities in the near East about 200-300 generations ago;

the *modern high-energy phase*, which began in Europe and North America about 7-8 generations ago. The technological developments made during this very recent phase have allowed a massive and completely unsustainable population explosion, along with huge increases in the rates of extraction and consumption of non-renewable resources.

Boyden's fourth phase may well be our last, unless humanity manages to make an unprecedented transition to the *restoration phase*, in which a more sustainable relationship between the human enterprise and the rest of the world is recovered. The question of whether, and if so how, humanity can make that last transition is central to all environmental activism.

The various outcomes of the interactions between genes and culture have been closely analysed by Durham (1991). He starts from the assumption that they cannot be independent, since behaviour and learning capacity may depend on genotype, while the genetic system may be modified by cultural selection. In principle, he identifies five possible forms of interactions.

Cultural mediation. In a genetically variable population, it is possible to plot the relative fitness of different genotypes in given environmental and cultural conditions. When the cultural traditions of a population change the relative fitnesses of different genotypes, then culture mediates natural selection of the favoured ones. For example, malaria is a devastating disease, caused by parasites located in the red blood cells, and sickle-cell anaemia is a widespread, genetically controlled disease, caused by a recessive allele which destroys the red blood cells (making them look 'sickle-shaped' under the microscope). In a population in which the proportion of genotypes having the sickle-cell trait ranges from 0-100%, an individual with one hundred per cent sickle-cells will die of anaemia, so the normal fitness of the trait is zero. But an individual with a mixture of sickle-cells and normal red blood cells has a much higher than normal resistance to malaria. In many parts of Africa, malaria is endemic, and there the individuals carrying some proportion of sickle-cell genes are at a great advantage. Cultural practices, such as slash-and-burn agriculture in tropical forests which greatly increases the number of breeding sites for malaria-carrying mosquitoes, shift the relative genetic fitnesses of the sickling and non-sickling genotypes.

Genetic mediation. Where there is a huge range of variation in a cultural property, of which only a subset are favoured for biological reasons, the distribution of the receptive genotypes will determine the distribution of the cultural trait. For example, all contemporary human languages describe colours, but none uses more than twenty-two combinations of colour terms, out of the two thousand and forty eight that are logically possible—because, apparently, only those are permitted by the neurophysiological architecture of the human eye and brain, and their linked colour-perception mechanisms.

Enhancement. Milk is a natural product for infant mammals, but adults tend to find it indigestible. The normal incidence of the genes producing lactase, the enzyme required for adult humans to digest milk, is only about twenty per cent, and drinking milk causes sickness in the other eighty per cent of adults. In areas with a long cultural history of dairy farming, the incidence of lactase among adults may reach greater than ninety per cent. Cultural acceptance of adult milk-drinking in dairy-farming populations has created an environment favouring natural selection for lactase, despite its initially off-putting effects. That in turn enhances the health and genetic fitness of the population.

Neutrality. If none of the possible variations in a cultural trait has any consequences for the genetic fitness of the population, all are free to develop unhindered by any biological considerations. This explains the riotous variation in cultural traditions of music, art and architecture, as opposed to the functional designs (not the surface decorations) of tools that are constrained by universal principles of efficient engineering.

Opposition. It is possible to find examples of cultural traditions which actually oppose the genetic fitness of the populations that practice them, but if the effect is severe, they will be short-lived. The example given by Durham is the epidemic of a devastating neural disease among the South Fore tribe of New Guinea, transmitted by cultural acceptance of cannibalism of deceased relatives. Examples from our own contemporary culture could include tobacco, hard drugs and drink-driving.

Modelling the interaction of genes and cultural traits allows exploration of questions, relevant to theology, that lie at the interface between culture and population genetics. The most important conclusion arising from the ideas of gene-culture co-evolution is that group selection of cultural traits can be stronger than natural selection of personal self-interest, and that process can and does favour attitudes that benefit the group at the expense of the individual (Wilson and Sober 1994). This is a welcome development, because the environmental crisis cannot be averted without some means of asserting the primacy of public over private interests; yet traditional analyses tackling genes and culture separately have so far denied that any such assertion is even possible.

4.4 Game theory and the social contract

Philosophers have recognised for centuries that all individual humans are capable of both good and evil. The problem is to determine the relationship between the two. Is one inborn and the other learned? Is it a matter of priority of one over the other—either that human nature is basically good unless corrupted by evil, or, alternatively, that it is basically evil unless redeemed by good? There have been plenty of people down the ages prepared to assert one or other of these positions. More recent studies show, however, that the good and evil sides of our nature are not to be so neatly separated. As Jesus saw so long ago, they are like the wheat and the tares growing inextricably together until the day of harvest (Mat 13:30), and the reason why is illustrated by modern research on game theory.

The difficulties of making people co-operate for the common good (what Grant (1993) calls 'the odds against altruism') were integral to the idea of the social contract, developed in the seventeenth century by early modern thinkers such as Hobbes and Locke. It was an imaginary device to explain how solitary individuals or families have come together to form a society, accepting obligations to one another and binding themselves to a sovereign able to enforce those obligations (Honderich 1995:163). Their philosophical point was (and is) that political authority depends on individual consent, that is, the consent that rational individuals would give if they ever experienced life without authoritative rule. To make this fictional consent plausible, the early modern political theorists had to tell a story, which rests on crucial assumptions about the asocial condition of humankind as it used to be before or without political authority. Like the Biblical myth of Eden, its purpose was to make a comment, not about human society as it once was, but as it is or should be now. In the ideal society, co-operation for the common benefit is an unambiguous good, and rampant selfishness is evil.

The need for a social contract, as preferable to the 'war of all against all' assumed by Hobbes, or the idea that 'the sheer dangers of anarchy had forced beings who were natural solitaries to make a reluctant bargain' (Midgley 1994: 110), are based on a series of spectacular misunderstandings of the lives, minds and social relationships of our human ancestors and of the sociable primates that preceded them. Yet the central myth of unconditional freedom persists in our culture. Some western philosophers continue to assert that it is self-evident that we are, above all, freedom-seeking creatures, even though most other cultures have not thought so, and despite evidence that does not fit, such as facts about our deeply social nature (Midgley 1994: 113). For example, Sartre and other existentialists placed great emphasis on the unique and radical freedom of human beings, to the extent that, 'apart from radical freedom, there is nothing whatever to say about the nature of humanity' (Rue 1994: 62). Would Hobbes or Sartre have considered these ideas as rational if they had been able to put them in the context of evolution and ecology? Darwin (who knew far less about primates than we do) thought not: he wrote in his notebook:[3] 'He who understands baboon would do more towards metaphysics than Locke'.

---

3. M notebook, p 84, 16 August 1838.

Game theory is a huge research field concerned with how and why people (and animals) make individual decisions in a social context. It has some features in common with political philosophy, but has been built up much more recently. Imported into biology from economics by John Maynard Smith, it has galvanised the study of animal behaviour. The classical, simple form of game theory, especially the famous Prisoner's Dilemma, also assumes that every decision is made *in vacuo* (I decide what I am going to do without knowing what you are going to do). In one-off Prisoner's Dilemma games a Hobbesian attitude does pay—the best thing to do is always to 'Look after Number One'.

The standard explanation of the Prisoner's Dilemma is a fable about two burglars, A and B, who were caught and questioned separately. The police try to get them to inform on each other, even though they have agreed between themselves to remain silent. The dilemma they each face can be explained by allocating scores to each possible decision. If A betrays B but B is silent, A gets off (score 5), and B is punished for both of them (score 0). If each betrays the other, both are punished (score 1 each). If both remain silent, the prosecution fails and both are freed (score 3 each). Obviously, the best thing to do is for both to remain silent, and both will score 3. But if A is silent and B confesses, then A will do time for them both (score zero) while B goes free (score 5). Remaining silent is a big risk for A, whatever B does. If both confess, the worst that can happen is that both get a score of 1, whereas each knows there is always the chance that the other will remain silent, in which case the one that spoke first will get the plum score of 5. Unless the commitment of each to the other is absolutely rock solid regardless of any threat or incentive the police might offer, the safest option for each in a one-off dilemma is *always* to betray the other, ie to make a totally individualistic decision, just as envisioned by Hobbes.

The Prisoner's Dilemma scenario does explain many short-term human interactions. People generally are wise not to trust vagrant traders that they may never see again, and co-operation between strangers meeting casually is often difficult to organise if there are no obvious barriers to a more direct me-first attitude. For example, in a bank, post office or supermarket with many service points, all customers will in fact get fair treatment (defined as gaining service in the order that they entered the premises) if they form a single queue, but they will not co-operate with each other to do this unless there is a barrier to prevent others from barging past the patient ones and getting to a window first. Banks generally do install such barriers now, but supermarket checkouts do not, and the competition to spot the fastest-moving queue cannot be eradicated.

Hardin's Tragedy of the Commons can be explained in terms of a classic Prisoner's Dilemma, by defining A as any given cattle herder and B as all the others (Boyd and Richerson 1985). The reward for mutual restraint (score 3) is unpredictable, and it carries risk that A might be one of the herders that is affected by the extra damage done to the pasture by someone else who does not share the benefit (so A gets the sucker's payoff, 0). The best possible reward available for A is for adding another cow (score 5), even though this reward is greatly reduced if everyone does the same and the pasture is damaged (score 1) than is the reward for mutual restraint (score 3). Under the (unrealistic) conditions specified by the standard, one-off game, in which all parties can be relied on to behave as individual egoists who are greedy and/or playing safe, all will betray the others and go out to the cattle market at once, and the result is inevitable system degradation.

On a wider scale, environmentalism in general is the same issue - it is a form of the Prisoner's Dilemma game involving many players, and the problem is how to prevent egoists producing pollution, waste and exhausted resources at the expense of more considerate citizens. So, says Ridley (1996: 225), we should accept that there is no innate tendency in humans which can be used to develop and teach restraint in environmental management. Far more than the social insects, humans who are expected to co-operate together socially are also in genetic competition with each other. Because we all instinctively watch our backs, *environmental ethics has to be taught in spite of human nature, not in concert with it*. Environmental activists and authors who eloquently describe the approaching global crisis, and plead for a fundamental shift in human values, underestimate the power of the games people play. At first sight, game theory seems to vindicate Hobbes: we don't co-operate because it doesn't pay us to do so unless everyone does, and only the most naïve would ever stake their all on the assumption that everyone will (p.42).

4.4.1 The importance of community life

Until very recently most human interactions have almost always been repeated interactions with family, friends and colleagues within long-term, stable communities. And repeated games of Prisoner's Dilemma between the same players give very different results from the standard one-off game–mainly because repeated interactions allow each round to be treated *as a deal rather than as a match*. The difference is that between a zero-sum game* such as football or tennis, in which only one side can win, and a non-zero-sum game such as social life generally, in which the more people co-operate together the more they all benefit. Real life is usually a non-zero-sum game, so the implications of this conclusion for social evolution are profound. At long last, it seems, science has discovered an empirical confirmation for what philosophers and moralists have always known, that the world is a better place when everyone restrains their personal self-interests. In terms of evolutionary psychology, the ethical systems of the world's major religions are (among many other things) comprehensive schemes of conscious instruction for maximising non-zero-sum social interactions.

The logic behind this conclusion was first explained to non-specialists by Robert Axelrod in *The Evolution of Co-operation* (Axelrod 1984). The age of computers has provided a powerful means of exploring more complex forms of game theory, such as variation in conditional strategies (what's best for me to do depends on what you did last time *and what you or someone like you might do next*). For example, which side of the road one drives on is a conditional strategy; the best side to choose is the left in Britain, but the right in France. Since neither side is intrinsically better than the other, the choice is determined solely by what everyone else does. It is true that British cars are configured to drive on the left and French cars on the right, but British tourists in France wisely take more account of the habits of French drivers than of the configuration of their own cars when deciding which side of the road to drive on in Paris.

To explore such conditional interactions, Axelrod organised a tournament for computer games, which showed that the best strategy to win a long series of many rounds of iterated Prisoner's Dilemma is to co-operate—provided three conditions are met:

1. The total number of games is unknown (defection is always the best strategy on the last move, and so the opponent would do best to defect on next-to-last, and so on back down the series). Hence, Axelrod says, one of the determining factors favouring co-operation is that the 'shadow of the future' should be long, and no-one knows whether they will be playing another round or not. *Translation: you never know your luck.*

2. There is an element of retaliation (if A tries defection, B should defect on the next move). *Translation: don't treat me like a doormat; completely unconditional love is not a practicable strategy for ordinary humans.*

3. There is an element of forgiveness (if A repents, B should co-operate again on next move). *Translation: I won't hold grudges if you don't.*

The best strategy, the one that won the tournament (and the next one as well) was operated by a programme called Tit-for-Tat. Its rule was very simple: always co-operate on the first move, and then do whatever the opponent did last time. This strategy was very stable in the long term, because it could not be exploited by more aggressive players (for fear of retaliation) but at the same time it was also forgiving, so when it got into a steady relationship with another co-operative player, it would never defect, and both benefited indefinitely.

There were disadvantages, of course. The first was that the total payoff earned by those playing Tit-for-Tat was smaller than could be gained by more ruthless players. On the other hand, the ruthless ones were seldom very long-lived. Unco-operative players did well at first—they tended to devour all the weaker, co-operative players and to gain a high score in a short time, but after a while only aggressive players were left, and these gradually eliminated each other. As Jesus pointed out, those who live by the sword will soon die by the sword (Mat 26:52). Paul added that those who hate and devour one another will be consumed by one another (Gal 5:15). In business and politics too, no-holds-barred aggressive strategies rarely last for long.

The second disadvantage was that Tit-for-Tat was also vulnerable to mistakes. If one player defected by accident, the other retaliated, and so did the first, and so on until they got bogged down in mutual recriminations. The histories of Northern Ireland and of Palestine are all-too familiar examples. Each successive round of revenge and counter-revenge could go on for months, and each increases the risk of really dangerous escalation. But on the other hand, if both partners are playing Tit-for-Tat, it takes only a single offer of reconciliation by one, and the other can immediately co-operate again. Provided both continue to do the same, both can step back from the brink and then survive indefinitely. So the problem of endless recriminations can be avoided by a simple adjustment to the programme, called Generous-Tit-for-Tat. This strategy can allow for occasional mistakes and does not retaliate at once—so that, if the opponent co-operates again, peace is restored. This is the only programme that can do better than Tit-for-Tat in a standard tournament. The implications of this adjustment for politics and social life generally—and the echoes of Gospel teaching—are obvious.

The conclusion seems to be that indefinite co-operation is possible so long as one has a high chance of finding co-operative partners and does not get too greedy. The question is, in a (human) world full of independent egoists as envisioned by Hobbes, how can Tit-for-Tat ever get started? The answer is, the world never was full of independent human egoists.

Hobbes' assumption that people make completely independent solitary decisions about social life was simply wrong. For the whole of the four to five million years or so that hominids have been evolving, and for some thirty million years before that during which the anthropoid ancestors of the human line were evolving, there has been no such thing as a solitary, totally independent individual, except maybe a dead one. Sociality always has been and still is as much a part of the definition of being human as is bipedal gait and a large brain, and it preceded both those characters by many millions of years. Not all primates are sociable, of course, but it is virtually certain that all species of humans and of their immediate ancestors, the australopithecines, always have been.

Within a stable social group, successful Prisoner's Dilemma strategies such as Tit-for-Tat can get their start, and they clearly demonstrate the advantages of local co-operation in a hostile world. Of course, there is always the danger of meeting someone playing a more hostile game outside the kinship group, and there are always real limits to, and conditions attached to, altruism in nature—but over the long term the advantages of getting frequent reinforcement from the locals are well worth the occasional ripoff from an outsider. Nevertheless, co-operation among unrelated egoists is a fragile thing and can evolve spontaneously only on certain conditions. In humans, a system of hard-to-fake signs of commitment (eg, as encouraged by hard-line religions) are important in identifying reliable partners in costly co-operative enterprises such as a political resistance movement, a business or even a marriage (Irons 1996). Even among primates, co-operation needs regular reinforcement by a predictable system of retaliatory justice. That is supplied in the first instance by social disapproval of cheaters (Boehm 1997), which then, in literate societies, becomes formulated into systems of written law (Irons 1996).

The literature on iterated Prisoner's Dilemma has dominated research on the evolution of co-operation between unrelated individuals, but there are problems. For example, the mathematics of games involving large groups of players, such as a whole tribe or nation—the normal human situation—soon become very complex, and their explanation by game theory alone may be questioned. Besides, there are other mechanisms. Two of them are potentially useful for the future of environmental activism; group selection (p.212) and the idea of co-operation as a by-product of emergency action in the face of a common enemy (Dugatkin and others 1992). The environmental crisis can certainly be classified as that.

4.5 Models of the origins of morality

Centuries of debate about the origin of ethics, observes EO Wilson (1998: 265-6), come down to this: either ethical precepts are independent of human experience, handed down to us from a transcendental source outside ourselves, or they are human inventions, open to empirical investigation. The choice between the two makes all the difference in the way we view ourselves as a species. It measures the authority of religion, and it determines the conduct of moral reasoning.

Every thoughtful person has an opinion on which is correct, but the split between them divides, not religious believers and atheists, but trancendentalists and empiricists. The question of the existence of God is another matter, and cuts the other way—a trancendentalist could believe in the independence of moral values, whether derived from God or not, and an empiricalist could believe in the human origin of moral values, whether guided by God or not.

Under Wilson's classification, I am a Christian empiricist. To me it seems clear that any character which, like sociality, has been ingrained in our nature fully as deeply and for much longer than our large brains must exert a powerful influence over our lives. And since morality is a key part of the problem of caring for creation, and sociality is necessarily linked to morality, we must pay it serious attention. If we wish to understand the processes that shape the human mind and spirit, we must first understand the processes that shaped the human species. As Mary Midgley puts it:

> once we accept our evolutionary history as a general background, it is quite natural and proper to use it in explaining many elements of human life. If we shut morality off from that explanatory pattern of thought, we tend to make its relation to the rest of human life unintelligible, which cannot be an advantage (Midgley 1994: 14).

Of course, that is not to say that what is natural is necessarily good. There is no need to adopt the ruthless values of natural selection as our own. But it is important to understand where our moral values come from, since if we modern people see a need to develop values different from those favoured in our ancestors by gene-culture co-evolution, or wish to change some disconcertingly stubborn parts of our nature such as the usual response to appeals from green activists ('what's in it for me?'), we need to know what we're up against (Wright 1994: 31). That is why Christians concerned to make a useful contribution to the environmental debate must understand what evolutionary psychology has to say about the evolution of public morality and about how community decisions are made.

Very many of what we like to think of as human characteristics have their roots in pre-human behavioural patterns, and to recognise them is not to humanise animals but to show what an enormous animal inheritance remains in humans. That does not mean that humans are *merely* 'naked apes', but it does mean that we need to understand the animal background which provides the historical context of our present moral dilemmas. Understandably enough, the classical philosophers and anthropologists had no access to that information, which explains why some of their assumptions are now having to be revised. For example, Elliott (1997) points out that no ethics can be grounded in biological impossibility, or allow behaviour that ends all further ethical behaviour. To the extent that the classical formulations were constructed on *a priori* human ideas such as equality and personal freedom, ignoring biological facts about, say, the social hierarchies natural among sociable primates and the finite limits to the earth's production capacity, to that extent those formulations must now be seen as contingent, not absolute—in Wilson's scheme, human inventions rather than independent laws of the universe. Because most human rights, laws and freedoms depend on the health and strength of the earth's ecosystems that support them, most cannot be universal, necessary and unconditional. Ultimately, facts can over-rule moral beliefs.

Classical philosophers such as Hobbes imagined that the default* setting of human nature is the solitary individual engaged in perpetual war against everyone else, and that was perhaps a reasonable idea in the state of knowledge about the natural world available to him. It is certainly a revealing comment about how he perceived the state of western civilisation in his time. But if he had been born into a hunter-gatherer society, he would certainly have known the huge extent of the mutual obligations and co-operation that bound the community together, the joint work of kin selection*, reciprocal altruism* and cultural indoctrination. Hobbes' philosophical understanding of the origin of ethics now has to be replaced with one based on biology, such as Richard Alexander's study *The Biology of Moral Systems* (1987).

Most biologists agree that biological processes similar to those that operate among animals have influenced human society from its origins, and have engendered fundamental human attitudes still taken for granted in the modern world. For example, the taboo on incest is practically universal, both in human and in animal societies. Almost all animal parents that care for their young will discriminate in favour of their own offspring and ignore others; likewise, any human parents who gave no more care and affection to their own children than they did to all others would be seen as monsters (Midgley 1994: 146). Conversely, the literary stereotype of the cruel step-parent is based in hard fact, both in human and animal societies. Male lions are certainly different from humans in the *extent* of their willingness to dispose of their predecessor's offspring (p.204), but not in *principle*. One careful survey found that pre-school children living in step-parent-natural-parent homes were *forty times* more likely to become abuse statistics than similar children living with two natural parents—largely because there are sections of human society in which small children of former unions are seen to detract from the remarriageability of custodial parents (Daly and Wilson 1997).

Various means of discriminating altruisms, plus early forms of cultural traditions, can all be observed in the living primates, so they must certainly have been well-developed among early humans. Hence, the default setting of human nature is not, as Hobbes believed, raw individualism: it is the need and ingrained habit of fitting in with, and looking out for, one's own closest group. The kinship bands of our ancestors were real communities, not the faceless crowds we encounter in modern cities; they were places where individuals could find their place in society and in the world (Rue 1994: 61). Even those at the bottom of the social heap benefited from sociality, because outside of their group, their chances of survival were drastically reduced, or nil. With increasing populations and resources, kinship bands could associate into tribal alliances, and then into kingdoms and states. The progress of this development among the Hebrews is well documented in the Bible:

> The Old Testament was the first genetics text of all. It
> is largely a record of separation: of who is among the
> chosen. The idea of universal relatedness, a common

humanity, is restricted to the New. Like most religious writings, both are codes for regulating society. Some people are labeled inferior, others are born to rule. In Biblical times, as now, human qualities were seen as innate and beyond control; the future, for good or ill, was set at birth. Kinship ruled those ancient lands, and, in spite of the supposed tolerance of the times in which we live, it retains its power today (Jones 1996: viii).

Even in egalitarian social groups of animals, the tradable 'currency' of favours regularly includes food or infant care, but social hierarchies open up additional possibilities. The connection between social status and breeding success makes status a resource in itself, so the exchange of status-enhancing favours is no different in principle from the exchange of food. Partners in such transactions develop coalitions, and coalitions can achieve much greater things in chimp, or human, society than any individual can do alone. The roots of manipulative human social behaviour and horse-trading of favours—such as that famously commended by Jesus in the parable of the unjust steward (Lk 16:1-8)—clearly run a lot deeper than most Christians would like to admit.

Throughout most of human evolution, the social environment to which natural and cultural selection gradually adapted our ancestors was one of overlapping primary and secondary groups, in which kin selection, reciprocity and group selection (in various combinations) worked reliably to ensure collective action, such as the sharing of food. But on the other hand, the more structured and socially cohesive a group, the more hostile it is likely to be to other groups. A flock of starlings is not a society in the sense that a band of chimps is, so starlings of different flocks simply ignore each other. Individual humans (and chimps) typically make great investments in their own social security and status, which helps to explain why humans, although among the most collaborative of all species, are also the most belligerent (Ridley 1996: 193). So the default setting of human nature, the force behind the moral attitudes that come most naturally, is to be conditionally co-operative with members of our own group, but much less co-operative with—at least wary of, and if necessary hostile to—members of other groups (Alexander 1987). Any serious study of the Christian attitude to the environmental crisis must examine how these processes work, and understand their implications for the efforts of green activists to stimulate a moral response.

Morality has always been considered to be a uniquely human achievement, but the precursors of it are well-documented in nature. That means that, strange as it may sound, morality preceded culture, and therefore religion, in evolutionary history: it is logically derived from evolutionary processes operating on the reciprocal behavior of intelligent creatures living in social groups (Reynolds and Tanner 1995: 14). Darwin himself predicted that it would be

> exceedingly likely that any animal whatever, endowed with well-marked social instincts, would inevitably acquire a moral sense or conscience, as soon as its intellectual powers had become as well-developed, or anything like as well-developed, as in man (Darwin 1871: 72).

Contemporary philosophers agree: hence Mary Midgley opens her book *The Ethical Primate* with the words:

> Human morality is not a brute anomaly in the world. Our moral freedom is not something biologically bizarre. No denial of the reality of ethics, nothing offensive to its dignity, follows from accepting our evolutionary origin. To the contrary, human moral capacities are just what could be expected to evolve when a highly social creature becomes intelligent enough to become aware of profound conflicts among its motives (Midgley 1994: 3).

Both intelligence and morality seem to be characters that could only have evolved among sociable animals, because they both facilitate the resolution of conflicts within and between sociable animals dependent on living in a group but also having different individual interests (de

97

Waal, 2000). In other words, the primary origin of the moral instincts is the dynamic relationship between co-operation and defection (Wilson 1998: 280).

In nature, conflicts of interest arise from the interactions of genetic rivals. Biologists familiar with the ideas of evolution have no trouble accepting that it is natural selection, acting on the immense reserves of genetic variation that every surviving species has evolved over generations, that has produced the huge variety of individual strategies for living that we can see in the natural world. But many people are uncomfortable with the implication that genes could have a determining influence on human behaviour, or that such explanations can cast any light on what ethicists have always believed was genuine altruism or truly moral behaviour.

The argument revolves around several separate but related questions. First, to what extent is human behaviour directly influenced by genes, in the way that is widely assumed to follow from Dawkins' famous selfish-gene metaphor, and are genes really 'selfish' anyway? Second, how could true human morality be derived by natural processes? Third, and most important in the present context, what are the implications of these ideas for Christian attitudes to creation? And do they help us understand human behaviour relevant to the present environmental dilemma?

4.5.1 Unconscious morality: the metaphor of the invisible hand

The genes that help to keep the fittest antelope one jump ahead of a cheetah are benefiting the antelope species as a whole as well as their own chances of replicating themselves into the next generation. But that was not their aim: they have no aim, and any beneficial consequences a gene might have for the species as a whole are purely incidental. It so happens that, in most living species, characters that favour individuals often do also benefit the group; if the balance changes, the species does not survive much longer. Adam Smith's famous image of 'the invisible hand' applied the same idea of the incidental benefits of self-interest to human economics in only slightly different words. If this process is widespread, maybe it could count as a form of unconscious morality, which might to some extent reduce the need for conscious morality.

But in human society, the successful pursuit of the self-interest of one person or company (if it is favoured by short-term cultural selection) certainly does not always benefit society as a whole—as New Zealand's recent political history amply illustrates. Did Adam Smith forsee this problem? Not according to Ormerod, who points out that Smith's economic philosophy *assumed* a moral climate in society which was generally accepted in his time but which we have largely abandoned. Smith saw

> the enlightened pursuit of self-interest . . . as the driving force of a successful economy, but in the context of a shared view of what constitutes reasonable behaviour . . . for Smith, . . . self-control was a natural, integral part of human behavior (Ormerod 1994: 13).

In other words, although Smith is taken by many to have advocated the ruthless free-market economics of today, he might in fact have been operating out of a paradigm more moralistic that that of the philosophers who criticise him, or the biologists who have imported his theories into current models of population genetics.

Whatever his intentions were, the fact remains that individual freedom is now a much more popular idea than public morality. That exaltation of the individual is part of the profound sea-change in western society generally labelled 'modernism', which Michael Northcott defines as 'the freeing of the spirit of individualism and capitalism from the traditional moral and ecological limits of ancestry and church' (Northcott 1994). In contemporary society, the conflicts of interest between self and beyond-self, private and public good, personal and group benefit, natural and cultural selection, are illustrated every evening on the TV news. Advertisers continually interrupt broadcast programmes in order to incite people to buy their products, whether they need them or not. Video shops persist in stocking dangerous films which degrade social relationships, give criminals ideas and permanently distort the growing minds of children—because the shop owners are free (within very broad limits) to select whatever stock most increases their profits, regardless of the social consequences. The anonymity of urban society allows modern thieves to opt out of the rules of reciprocal altruism that governed our ancestors, without being denied the rewards of social life, including police protection of their own property against other thieves.

In civilised societies, natural selection of personal benefit is restrained to variable extents by cultural selection through education and social facilitation. When the cultural restraints weaken, individual self-interest reasserts itself, as William Golding pointed out in *Lord of the Flies*. Real life dramas in any disaster zone play out the same theme. Amid the total breakdown of civilian authority in Somalia, Bosnia, Ruanda and Kosovo, armed gangs ruled by the law of the jungle; in the aftermaths of the Floridan hurricane and the Rabaoul volcanic eruptions, looters cynically ransacked the damaged shops and houses. Even in ordinary life, the first reaction of most people to any economic news (say, a change in the state pension scheme, or in the rules of taxation, or a rise in interest rates, or a revision of the laws governing working conditions or health care, or an international stock market crash) is to ask 'How are these changes going to affect my interests?' It seems to be as automatic and natural as breathing, one of our built-in survival mechanisms. It is easy to assume that we have such reactions because their survival value has been proved over thousands of years: any tendencies to ignore possible threats to personal survival were eliminated during our remote past. That implies that the metaphor of the invisible hand is mere wishful thinking. Is this true?

Proponents of sociobiology[*], launched by EO Wilson's massive book of the same name (Wilson 1975) do not hesitate to answer in the affirmative. Sociobiology has generated vigorous controversies for the last twenty years, by exploring the implications of extending gene-centred evolutionary theory to human behaviour. The row has given new life to the old nature-nurture debate, the question of whether inheritance or environment (natural versus cultural selection, again) most strongly controls individual development. The arguments on either side have been intense, and often unproductive. Philosophers such as Mary Midgley complain that

People have been strangely determined to take genetic and social explanations as *alternatives* instead of using them to complete each other. Combining them without talking nonsense is therefore by now fearfully hard work (Midgley 1978: xviii).

She goes on to argue that

the antithesis between nature and nurture is quite false and unhelpful . . . most activities of higher animals [are] both innate *and* learned' (*ibid*, 54).

Those who continue to study sociobiological ideas but do not wish to be associated with the political arguments it stirred up tend now to refer to their work as 'evolutionary psychology'. Both forms of the discipline are interested in modelling the genetic effects of specific behaviours, but there is a spectrum of opinions on how, and to what extent, genetic and cultural effects are linked. In turn, interpretations representing the 'strong' theories of gene-culture interaction have implications for ideas about the origins and functions of morality different from those of the 'weak' theories.

4.5.2 The metaphor of the selfish gene

Sociobiologists deal first in abstract models, mostly those exploring the workings of gene-centred theories of evolution developed by population geneticists, and only later—if at all—in real animals. Any act of what might appear to be unselfish generosity, such as sharing food, they call 'selfish' from the gene's point of view if it ensures that the donor will be given food or sexual favours in exchange. These in turn may lead to evolutionary advantage in terms of enhanced reproductive success. Altruism in animals, called by Wilson (1975) the 'central problem of sociobiology', is thereby explained in genetic terms, and in the natural world this insight has proved extraordinarily fruitful.

When applied to the higher animals that are also profoundly influenced by non-genetic, cultural sources of information, sociobiological models have to ignore cultural selection and the various other genetic processes besides natural selection that also influence the relative chances of a gene being transmitted to the next generation. Like the economic models with which they have so much in common, genetic models treat these effects as externalities. The danger is that mathematical geneticists easily become so impressed by the elegance and predictive power of their models, that they come to see the models as the only reality and the real world as a product of confused perception (Ward 1996: 28). Not only theologians and philosophers, but also practising geneticists, reject the selfish-gene metaphor as nonsense.

To be fair, we should note that it is possible to compile a useful list of the less controversial sociobiological insights, which are important and should be acknowledged (Cavanaugh 1996):

> Understanding human nature is easier if we understand biology.
> Culture is an elaboration of biology.
> Free will must be exercised within biological constraints.
> Moral systems have a biological component.
> Our security and even our happiness depend on living consistently within our biological natures.
> Our social structures—government, education, economic institutions, and religion—work best when they take account of our evolved social natures.

On the other hand, hardline reductionism, ie, attempting to interpret all the glorious complexity of the natural world as the unconscious product of natural selection operating at the level of the gene, is widely and severely criticised. Dawkins uses vivid language that often gives this impression, when he compares humans with 'robots' whose genes have 'created us, body and mind' (Dawkins 1989: 19-20). Critics reply that, whether or not it applies to animals, it is possible to apply this view to humans only when all the positive effects of cultural selection are ignored. From there it is a short step to genetic determinism and all its intolerant bedfellows. According to the deterministic view, people of

102

different races, genders and sexual orientation are born different, and there is nothing to be done about the inevitable disparities in wealth and status between them. Determinism can be a disastrous weapon in the wrong political hands.

Dawkins objects to this interpretation, pointing out that 'genes do not control their creations in the strong sense criticized as "determinism". We . . . defy them every time we use contraceptives' (*ibid*, 271). He reminds us of the well-recognised difference between the way a thing is and the way it should be. He asserts that his concept of the selfish gene is merely a hypothesis about our nature, constructed on the grounds that a prudent warrior takes thought to know his enemy before compiling any battle plan. He does not *approve* of selfish nature, as some of his opponents have tried to claim: quite the contrary. But he believes that genes and culture evolve independently, so any code of public morality we choose to develop must be done by deliberate, heroic choice against all the subtle influences of our genes. Such exhortations tend not to sound convincing when compared with Dawkins' frequent references to 'our masters, the genes'.

Even Dawkins' opponents agree that he does not argue that that genes alone determine behaviour, since

> that view is incoherent and no one holds it. [But there is] evidence for the view that they have *some* effect on behaviour, that can be found in a vast range of activities which cannot sensibly and economically be explained on any other assumption (Midgley 1978: 67).

Of course there are no *specific* genes for aggression or racial predjudice or general nastiness, any more than there are genes for legs or eyes. All complex structures and behaviours are produced by the sequential interactions of many genes acting together within bodies capable of a great range of alternative paths of development.

Considering the numerous criticisms of selfish-gene thinking, it is surprising that it has survived as long as it has. Stephen Jay Gould suggests an answer:

> Biological determinism has always been used to
> defend existing social arrangements as biologically
> inevitable—from 'for ye have the poor always with
> you' to nineteenth-century imperialism to modern
> sexism . . . the crude versions of past centuries
> [concerning, for example, the inferior intellect of black
> people] have been conclusively disproved, and its
> continued popularity is a function of social prejudice
> among those who benefit most from the status quo
> (Gould 1977: 258).

Gould's quotation from Jesus about the poor (Matt 26:11) completely misses the point Jesus was making, but his explanation for the popularity of biological determinism in contemporary times is astute, and, in my opinion, largely true. But the concept of biological determinism relies in turn upon Dawkins' 'gene's eye view' of evolution, and the intellectual credibility of both has been severely questioned in recent years.

Objection 1. Critics point out that the idea of the 'selfish gene' is only a metaphor, since genes have neither a self nor the emotions to make them selfish. This is true, although there is nothing wrong with that: metaphors abound in science, and in all other systems of thought that deal in abstract entities including theology (p. 25). But Dawkins has been fiercely criticised for persistently using, to general audiences, the word 'altruism' in its restricted technical sense, even though people invariably understand it only in its well-recognised but quite different meaning in common speech (Midgley 1979). In the process, he and other evolutionists have changed the defining criterion from *motives* to *effects*, and largely without comment (Wilson 1992). For some critics, this is evidence that Dawkins' metaphor of the selfish gene has become far more than a metaphor, a mere linguistic device; it is the 'formative vision' (that is, the paradigm)

> that shapes his fundamental understanding of life . . .
> the root metaphor is not the selfish gene, but
> selfishness as such . . . [this has] characterised modern

thought since Descartes and modern institutions since the emergence of free-for-all economics (Grant 1986: 446).

Darwin was himself a gentle person, but he was greatly influenced by the intellectual climate he grew up in—a ruthless industrial society in which merely staying alive was a struggle and only the fittest survived. Grant's thesis is that the origins of Darwinism as a way of understanding nature and (in its social versions) human nature were actually the opposite of what they are usually taken to be; rather than a theory of nature applied to human society, the theory itself reflected human life in the nineteenth century (Grant 1993: 105). As Kühn pointed out, the metaphysical assumptions of a culture exert a strong influence over the character of the scientific paradigms developed within that culture (Barbour 1997: 146). That is the reason why proponents of green ethics must frame their exhortations with reference to the interactions between economic ideas and science in western culture (*Figure 2*).

Objection 2. Critics further point out that sociobiology ignores an absolutely vital distinction between two very different forms of selfishness. *Evolutionary egoism*[*], self-service in the evolutionary sense, the operation of natural selection on unconscious genetic self-interest that is a property of genetic lineages, is not at all the same thing as *vernacular egoism*[*], or personal selfishness, the conscious individual self-interest that is a property of individuals and operates in opposition to cultural selection (de Waal 1996: 15). Behaviour can be personally selfish but genetically altruistic, and vice versa (Wilson 1992). The application of economics-based models to estimate the inter-generational benefits of natural selection on genetic self-interest in, say, the evolution of parental investment in animals, is often appropriate and useful; but the personification of isolated genes as independent active agents capable of personal selfishness (Bowker 1995) is certainly not. Dawkins claims that he does try to avoid talking in such terms, but although he might understand the distinction, many who read him do not—and Grant's point was that to Dawkins the distinction is barely visible anyway.

Midgley (1978) does not deny the many valuable insights into human nature offered by the development of sociobiology, but she rejects on principle that it is even possible to apply them to human behaviour without any reference to motive (mention of motive is regarded as teleological, associated with Aristotle and rigorously excluded from sociobiology and all other branches of contemporary biology, at least in theory). For example, on the central problem of altruism, Midgley points out that two totally distinct problems become confused when observations from animals are applied to humans. For animals, the question is whether a particular behavioural trait can survive if it leads its bearer to do things which do not *in fact* pay it (or its relatives), whereas the same question applied to people becomes

> Can a conscious agent deliberately choose to do things that *he thinks* will not pay him? This problem can be considered only by people willing to take motives seriously . . . Officially . . . [sociobiology] ignores motives, but in fact makes constant reference to them, and because this reference is unacknowledged, its errors go uncorrected . . . [we must indeed] make full use of the evolutionary perspective as a background. But it is equally necessary . . . to be capable of dealing with the foreground—of abandoning the long perspective and looking directly at the motives of individuals. We must take these motives seriously in their own right and not try to reduce them to . . . behaviour patterns; . . . motives have . . . their own evolutionary history . . . and their own internal point, and it is virtually never a wish to bring about some evolutionary event, such as the maximisation of one's own progeny (*ibid*, 117, 128,142].

Midgley here identifies the problem that arises (again) when people fail to see the differences between natural and cultural selection. Natural selection does operate largely as a genetic filter, unconciously preserving in us the genetic bases of the individual attitudes that have been adaptive in the distant past. Cultural selection operates as an accumulator of memes (ideas), affecting groups as well as individuals,

and involving conscious motives and moral agency over the short term
arena of shifting cultural rewards. Human behaviour is influenced by
both; as Midgely points out, we need to understand the evolutionary
background, but that is not enough (*Table 2*). No account of human
behaviour can ignore motives, or reduce moral reasoning to differential
reproductive success. It is necessary to take seriously the point made by
multi-level[*] selection theory, that the experienced world does not only
have value if it serves Darwinian fitness; it has developed emergent
values of its own.

   Objection 3. The selfish-gene model is based on what Whitehead
(1927: 64) called the 'fallacy of misplaced concreteness'[*], the tendency to
organise knowledge in terms of abstractions and then to reach
conclusions and apply them to the real world as if abstractions and
reality were the same thing. But when the distinctions between
abstraction and reality are forgotten, trouble follows. Abstract models
cannot be all-inclusive: certain elements have to be omitted from them,
because, say, they cannot be defined precisely enough, even if they are
vital to the workings of the system modelled. Economists tend to label
these elements 'externalities', largely so as to avoid any challenge to the
model. Hardin's *Tragedy of the Commons* is an especially clear case of the
inexorable workings of uncontrolled externalities and the false
conclusions to which such thinking leads (p.47). Yet theoretical
economists continue to think this way, and that represents an extremely
serious danger to all the rest of us who have to live in a real world
governed by their experiments. The consequences for New Zealand are
eloquently described by Kelsey (1997).

   The fallacy of misplaced concreteness is a constant danger for those
who forget the distinction the between models based on gene-centred
evolution (interpreted through the metaphor of the selfish gene) and the
reality of whole, functional animals. The theory of gene-centred
evolution is based on extensive mathematical modelling, but models
necessarily have to simplify reality merely to make the calculations
possible. Vital features of real genetic systems, such as the hierarchical
and essentially co-operative nature of genes (Maynard Smith and
Szathmary 1995) are forgotten, and the very many other dynamic and
multi-level processes that determine genetic change are ignored (Wilson

1997a). So, although gene-centred evolutionary theory has had some astonishing successes in explaining animal behaviour in the wild, the models it is based on involve many simplifying assumptions.

The fallacy of misplaced concreteness lurks behind Dawkins' curious notion that, since genes exist in millions of identical copies unaffected by their transition from one generation to the next, they must be the effectively immortal engineers that design the all too mortal, temporary individuals they live in, and they are therefore by implication more important. Ward (1996: 137) parodies this idea by picking up Dawkins' own analogy between genes and cake recipes—as if, Ward says, the only point of a recipe (a *codical* reality) is to be replicated in cookery books, and the cakes themselves (as *material* realities) are unintended by products! The parallel is valid in so far as genes and recipes are both codes of information, rather than concrete entities such as cakes and bodies. The fallacy arises from the fact that codes of information and material bodies are entities belonging to different, mutually exclusive domains (Williams 1992).

Both are real and both are important, but the relative permanence of the genes does not make them more "important" than the temporary bodies they build—it is the *interaction* between them that matters. A reality in the codical domain is only *potential* experience, not real life. The modern printed score of a piece written by, say, Bach in 1710 is one of many identical copies of a musical code that has been reproduced unchanged for almost three hundred years, which makes the code more permanent than the generations of different musicians who have read it, but it is only *potential* music. It takes a (relatively) short-lived material musician (more significantly, an orchestra of individually variable and distinct cooperating musicians) to lift the code off the page and into a single unique performance experienced by the particular audience present on that night. The idea that the score could be important than the performance simply because it lives longer is absurd.

RC Lewontin, himself an experienced geneticist, ridicules the assumption that the gene determines the individual and the individual determines society. Despite the name *socio*biology, he says, the deeper ideology beneath it is the priority of the individual over the collective. The metaphor of the selfish gene is appropriate only to the extent that the biology of an individual is determined by its genetic make-up. That extent is greater in, say, a beetle than in a lion, but it is never one hundred per cent in any animal, and least of all in primates.

The underlying assumption is the one popularised by the film *Jurassic Park*, that if we had a large enough computer and knew the entire DNA sequence of an animal, we could construct that animal artificially. But that is simply not true: not even the animal computes itself only from its DNA. Any living animal at any moment in its life is the unique consequence of its developmental history, itself the interaction of genes and environment. It is not that the whole is more than the sum of its parts. It is that the properties of the parts cannot be understood *except* in the context of the whole. History transcends the limitations of either genes or environment to control us, because they have been replaced by an entirely new level of causation, that of social interaction (Lewontin 1991). In other words, by the emergent processes characteristic of the next level of selection.

4.5.3 Natural morality as self-deception

Evolutionary psychology is like sociobiology in that it recognises morality as a product of natural selection, just as is any physical feature. Wright (1994: 26) points out that the similarity in physique that makes every page of *Gray's Anatomy* applicable to all humans of all races applies also to their mental architecture—the basic structure of the human mind is species-typical. It is therefore reasonable to speak of 'the psychic unity of humankind'. Among the products of the evolutionary heritage of humankind is the limbic system that controls our emotions (Wilson 1975: 6), which stimulates the conscious feelings of love, fear, racial hatred, sexual jealousy and many more that profoundly influence our daily decisions and which are common to all people. The problem is that some of these run counter to traditional moral wisdom, which introduces severe personal conflicts, because over-ruling our deep-seated natural emotions is never easy. Freud knew that well enough, but he was wrong in his speculations that 'primitive man was better off knowing no restrictions of instinct'. As Wright (1994: 323) points out, this is a mere legend. It has been a long, long time since any 'primitive man' could enjoy 'no restrictions' on these 'instincts'. Repression and the unconscious are the products of evolution too, and were well developed long before civilisation further complicated human mental life.

Because all humans are absolutely and necessarily sociable, these emotions generally arise in social contexts and have strong social effects, both for good and bad. One of their most important functions is to facilitate reciprocal altruism. Feelings of gratitude encourage the return of favours within a group; callous indifference to suffering is part of the retributive impulse; both help to reward co-operators and discourage cheats, and so are among the strong governors of reciprocal altruism, hence both were advantageous for thousands of generations of our ancestors' lives. Both maintain their place in our lives by creating powerful feelings of righteousness in those that practice them, plus the mechanisms of social approval and disapproval that monitor the behaviour of group members, often by unconscious mechanisms designed to make us think we are better than we are (Alexander 1987: 139). All emotions that help to bind a group together and distinguish it from other groups are strongly favoured by cultural selection.

There is a darker side to this process as well. Tribal loyalty feels right because it helps to keep groups distinct, and therefore has always been strongly reinforced by cultural selection. Unfortunately, it quickly becomes racism, whilst retaining the feeling of rightness—so it is perfectly natural and socially acceptable to despise the enemies of one's own group, as the Jews despised the Samaritans. It was the anti-racist implications of the parable of the Good Samaritan that shocked the Jews who first heard it, not the generally accepted teaching that one should help a fellow-human in distress. In our overcrowded civilised world, racist feelings, which may still feel right and natural, are capable of even greater harm than ever.

At first sight, the idea that emotions drive moral behaviour is counter-intuitive. After all, the emotions are themselves the product of the same processes of natural selection that favour genetic self-interest in all life. Yet there is plenty of evidence to show that in the highest primates some emotions at least have become transformed into agents of social cohesion—which, combined with reflective, self-conscious intelligence, often involves the disciplining of personal self-interest and the beginnings of moral behaviour (de Waal, 1989). But how can such an effect be produced by natural selection? The answer suggested by one school of evolutionary psychologists is that the positive effects of natural, evolved morality work largely through the unconscious. Feelings of personal moral worth serve to mask the self-deception that is necessary to equate natural emotions with socially 'correct' behaviour. After all, self-righteousness could work as a moral force only if the self-interested basis of it is disguised.

But what about the argument that scuttles most forms of reference to evolution as an explanation of human affairs, the one based on philosphy rather than biology: any appeal to nature as a moral authority confuses what *is* with what *ought to be*. These are definitely not the same thing at all. But the supporters of the morality-as-deception argument reply that, under their scheme, the point is rather of *learning to be sceptical about our natural prejudices and intuitions*. For example, cynical indifference to the suffering of morally or socially unacceptable groups, such as gay AIDS patients, is an expression of the built-in biases of human nature, and is regrettably more common among religious fundamentalists than atheists. Loyal Rue (1998) calls this our 'default morality'. Contrariwise, logic and trans-ethnic religious teachings meet in the idea of the extension of reciprocal altruism to beyond the kinship group—to include not only other human groups but also animals and the whole material creation, as recommended for other reasons by Singer (1983).

On the other hand, a convincing argument against the evolved morality-as-self-deception theory has been put up by Francisco Ayala. He points out that it is only the *capacity* to construct moral codes that is rooted in biological evolution (Ayala 1998). Ethical behaviour in all its various forms depends on three necessary and jointly sufficient conditions, supplied by the biologically determined constitution of the human mind: (1) the ability to anticipate the consequences of one's own actions; (2) the ability to make value judgements; and (3) the ability to choose between alternative possible courses of action. The moral codes themselves are the product of cultural evolution—which explains why they differ so much between societies. It was intelligence, not the hidden machinations of our genes, that allowed humans to develop those codes in historic time. Could Ayala's interpretation offer some hope that intelligence might get us out of the real and present danger presented to humanity by the environmental crisis? Only if we ignore the lessons of human history, which show clearly that people do not always choose to do the right thing even when they know very well what it is.

4.5.4 True human morality: a rebellion against nature or the fulfilment of nature?

If humans are a product of natural selection, a process described by Williams as 'morally unacceptable . . . evil [and] abysmally stupid' (Williams 1996: 156), and the evolutionary roots of human ethics involve self-deception, where does true human morality come from, if it is possible at all? Within the empiricist camp, the possible answers to these questions form two schools of thought.

The first group, led by Alexander, Dawkins and Wright, picture a voluntary, deliberate sort of human morality, one that pits us as rational beings against the brute nature that not only formed us but also deceives us as to what it has been doing. Arnhart (1998) labels them 'Hobbesian Darwinians', who assume that humans are ineradicably self-centred. According to them, evolutionary psychology shows us how to look behind our evolved moral feelings and see the self-interested genetic machinery that drives them. We can then choose whether or not to obey. Effectively, they claim that we have outgrown our genes, learned how to examine them objectively and figure out how they operate—even that we are the only organisms on the planet capable of defying them. How can we do that?

One suggestion draws a parallel with the history of the upper chamber of the British parliament, the House of Lords. That venerable institution destroyed its own power to hold back the political development of Britain by assenting to the successive Reform Acts of 1832, 1867 and 1884, which vested the ultimate authority of Parliament in the House of Commons. In Lewontin's metaphor (1991: 123), 'the genes, in making possible the development of human consciousness, have surrendered their power both to determine the individual and its environment'. The idea is all the more remarkable because, as Wright (1994) put it: natural selection, a process devoted to selfishness, has eventually produced organisms which, having finally discerned its creator, could reflect on this central value and reject it.

To Alexander, Williams, Huxley, Dawkins, and their followers, Lewontin's metaphor is simply nonsense. The genes have not surrendered control, they have merely gone underground. Raw nature (as represented by the evolved form of self-deceiving morality produced by natural selection) is something that truly moral humans have to fight *against*, and true human morality is a counter-force, a rebellion against our brutish makeup, rather than an integrated part of human nature. In Rue's terminology, our 'default morality' produced by natural selection has to be modified by the moral code programmed into us by our culture, our 'over-ride morality', which takes constant effort and moral discourse, but is necessary if we are to find ways of superceding our biology:

> Chimps cannot manage anything as global as universal brotherhood. Nor could we, if we did not have the mediation of symbols to help us over-ride our default morality (Rue 1998: 533).

In other words, there is no such thing as natural altruism, and true human morality has to act *in opposition* to our animal nature, be imposed upon it in order to bind and control it. The classic quote is from Richard Dawkins, who believes passionately that if we are ever going to develop real unselfishness and a code of morals, we must do so by our own effort, because our genes will not help us:

> Be warned that if you wish, as I do, to build a society in which individuals co-operate generously and unselfishly towards a common good, you can expect little help from biological nature. Let us try to teach generosity and altruism, because we are born selfish ... We, alone on earth, can rebel against the tyranny of our selfish replicators [genes] (Dawkins 1989: 3).

The advantage of evolutionary psychology, according to its proponents, is that, the better we understand it the more likely we are to be capable of detecting and rejecting ancient prejudices. We do not have to take our cues from our primate origins now that we can see them for what they are: but on the other hand, if we do not see them we remain in danger of being manipulated by them.

But surely, a theologian would answer, all such ancient prejudices have always been perfectly well understood by traditional spirituality, and labelled with the single, simple and challenging word, 'sin'. Taken seriously and understood from the heart, existing Christian teachings and sacraments already offer more than enough resources to anyone genuinely concerned to deal with them.

The second group of empiricists, represented by de Waal and Gould, is labelled by Arnhart (1998) the 'Aristotelian Darwinists', because they disagree that the conscious motives that underpin human morality must necessarily be independent of, and superimposed upon, the self-serving evolutionary process that shaped our minds and bodies. They dispute the reductionist view of humans as potentially but not naturally moral, as hypocrites living in constant denial of our thoroughly selfish nature (Wright 1994). For example, Gould points out that

> Basic human kindness may be as 'animal' as human nastiness . . . our genetic makeup permits a wide range of behaviours—from Ebenezer Scrooge before to Ebenezer Scrooge after . . . Functioning societies may require reciprocal altruism. But these acts need not be coded into our consciousness by genes; they may be inculcated equally well by learning (Gould 1977: 266, 257).

de Waal (1996; 2000) provides plenty of evidence from primatology to support Gould's assertion that limited forms of kindness and co-operation are natural to the sociable animals nearest to us. These actions convey a direct benefit to individuals, not merely to their genes. Among our closest animal cousins, social environment shapes and constrains individual behaviour. Evolution *has* produced the pre-requisites for true human morality, including mechanisms to defuse tensions within a group, a tendency to develop social norms and enforce them, the capacity for mutual aid, a sense of fairness, and so on. All these are regularly observed in captive chimpanzees, and the comparable processes in us are merely extensions made possible by our higher intelligence (de Waal 1996):

> The fact that the moral sense goes so far back in evolutionary history that other species show signs of it plants morality firmly near the centre of our much maligned nature. It is neither a recent innovation nor a thin layer that covers a beastly and selfish makeup (*ibid*, 218) . . . Instead of concluding that morality is a cultural construct that flies in the face of nature, Huxley and his followers would have done better to broaden their perspective on what the evolutionary process can accomplish (162) . . . To give the human conscience a comfortable place within Darwin's theory without reducing human feelings and motives to a complete travesty is one of the greatest challenges to biology today (117).

Ridley (1996: 141, 144) agrees:

> Virtue is indeed a grace . . . something to be . . . drawn on and cherished. It is not something that has to be created against the grain of human nature—as it would be if we were pigeons, say, or rats with no social machine to oil. It is the instinctive and useful lubricant that is part of our natures. So instead of trying to arrange human institutions in such a way as to reduce human selfishness, perhaps we should be arranging them in such a way as to bring out human virtue.

In other words, 'grace does not destroy nature, but *perfects* it', as Aquinas maintained (Hall 1986: 138). In terms of the gene-culture co-evolution models of Durham (1991), this is a case of enhancement (p.84). Cultural traditions such as religion reinforce existing group selection favouring the better survival of the most strongly co-operative social groups. It is our best hope for the future.

4.5.5 Religion and morality

I see the debate between the Hobbesian and Aristotelian Darwinists as the old Pelagian argument in modern dress. The Hobbesians represent Augustine's position, that human free will is weakened and incapacitated by our inherited nature - damaged not by Adam's sin, but by the self-deceptive mechanisms built into the animal foundations of our moral systems. All the moral outrage characteristic of Augustine's disparaging view of the origins of human moral conflict reappears in the writings of Dawkins and Williams. On the other hand, the Aristotelians represent Pelagius' position, that God made humanity knowing precisely what it is capable of doing (McGrath 1994: 373); moreover, humans retain the rationality and free will that God gave them, so in theory are capable of making real moral decisions. De Waal's more charitable view of the origins of moral conflict presents a real contrast to that of the Hobbesians. The crucial repercussions of the historic Pelagian controversy for the Christian doctrines of grace and merit have no modern parallel, since they are not relevant to the Darwinists' debates, but the difference in the tone of their attitudes to human nature is comparable. For de Waal (1996: 17), the Hobbesian's proposed abyss between morality and nature is quintessentially Calvinist—the age-old half-brute, half-angel view of humanity.

One of the puzzles of this debate is that the proponents of the idea of morality as a rebellion against selfish nature might have been expected to acknowledge religions as at least potential allies. The key selective value of early religions was that they reinforced and fostered group-centred altruism (Barbour 1997: 263; Burhoe 1979; Wilson 1977). Of course there is also competition for cultural and biological success *within* the group (Irons 1997), but that is temporarily set aside when the group is faced with a threat from outside.

On the contrary, the reductionists see *both* morality and spirituality as evolved, unconscious self-deceptive mechanisms now embarrassingly revealed by the cold light of scientific logic. They allow that spirituality might once have been an important cultural arbiter of morality, but that time is now past (Wilson 1977). The traditional religions are seen as part of the evolved but now outdated animal heritage that truly enlightened humanity has to fight against:

> traditional religious beliefs have been eroded, not so much by humiliating disproofs of their mythologies as by the growing awareness that beliefs are really enabling mechanisms for survival (Wilson 1977: 3.)

To the hard-line reductionists, the process of looking behind our evolved moral feelings and seeing the self-interested genetic machinery that drives them therefore requires the rejection of all traditional religion. In so far as religious behaviour is capable of fostering inter-group rivalry, even to the extent of causing local extinction of the population (as on Easter island: Bahn and Flenley 1992), the reductionists' bitter criticism of religion does have a point. In terms of the gene-culture co-evolution models proposed by Durham (1991), their understanding of true human morality interpreted as a conscious rebellion against a genetically-based system of self-deception could be seen as a case of opposition (p. 85).

However, not everyone would agree that this is a fair or accurate description of religious behaviour. For a start, the level of religious belief among scientists in general, although not a majority, is still substantial (about forty per cent) and has not changed much over the last eighty odd years (Larson and Witham 1997). Furthermore, in so far as religion can encourage group-centred, cultural-based altruism at the expense of personal selfishness, which in turn favours the genetic survival of the group, then religious traditions can be a cultural force which favourably influences the direction and/or strength of biological evolution. To those who believe that the survival and prosperity of their societies are more important than their own, a functional explanation for religious beliefs does not undermine them at all–rather, it constitutes an immediate reason for accepting them (Austin 1980: 194), whether or not that acceptance is accompanied by any real conviction about the existence of God. Such an attitude requires the future, thinking church to accept as valid a rather deliberate sort of faith, but not one to be sniffed at:

> In the past those who came to see that religion is just human became themselves non-religious. Today this is no longer the case. The first *conscious* believers are appearing, people who know that religion is just human but have come to see that it is no less vital to us for that (Cupitt 1984: 19).

In terms of Durham's models, this interpretation of the role of religion would be better seen as a case of cultural mediation (p. 83).

Cultural mediation of genetic fitness via religious practices could reasonably be regarded as a good thing, so the question arises: what are the implications for Christian theology if the calculus of the genes is there underneath, cloaked by an evolved, unconscious moral sense that prompts us to behave nobly anyway—at least towards members of our own group?

I agree with de Waal that there is no problem here provided we are able to distinguish between process and outcome. We do not need to let the ruthlessness of natural selection distract us from the wonders that it has produced. Just as diamonds are the product of intense heat and pressure, or birds and aeroplanes appear to defy the law of gravity yet are fully subject to it, moral decency may appear to fly in the face of natural selection yet still be one of its many products (de Waal 1996: 16, 12). We need not be ashamed of the genetic processes that formed us, like children embarrassed by their old folk, but neither need we be dependent on them now.

> Humans and other animals have been endowed with a capacity for genuine love, sympathy, and care—a fact

that can and will one day be fully reconciled with the
idea that genetic self-promotion drives the
evolutionary process (*ibid*, 17).

If, as Christians believe, God is involved in evolution from the very
beginnings of life until now, it is reasonable to expect God's work to be
all of a piece, that the thrust will be all in the same direction. If morality
is the highest capacity of free, conscious creatures capable of entering
into a spiritual relationship with God, then it may be expected to *fulfil*
nature, not to combat it. I see no difficulty with Wilson's (1975: 120)
interpretation of spirituality as just one more Darwinian enabling device
somewhat like nursemaids, nurturing our ancestors as they stumbled
along the slow road from *Australopithecus* to Christ, *provided* we do not
remain as infants in need of nursemaiding indefinitely. We have to grow
up, through all the stages of evolution from natural selection though
cultural selection to God's realm of no selection (Theissen 1984). In
Paul's famous analogy,

> When I was a child, I spoke like a child, I thought like
> a child, I reasoned like a child; when I became a man, I
> gave up childish ways (1 Cor 13:11).

Wilson's suggestion is valid when taken to refer to tribal religions,
but not when applied to fully-developed Christian theology:

> All our highest ideals and most difficult aspirations,
> the agonies of spirit and the heroisms of moral
> commitment, stand revealed as mechanisms for the
> multiplication of DNA. The suggestion is so
> extraordinary that we might wonder if any rational
> person could make it seriously . . . what the biological
> moralists are really doing is to impose their own moral
> ideals on the evolutionary process . . . They reduce the
> aims of morality to the minimal end of survival, thus
> depriving survival of any particular point or purpose

beyond itself . . . and [they] reduce the importance of moral endeavour and individual responsibility virtually to zero (Ward 1992: 67,77).

After quoting Dawkins' call for conscious altruism to fight against 'the tyranny of our selfish replicators', Ward asks

> Why should he *hope* that altruism is possible, unless he really does have a basic sense of moral obligation? And isnt it odd to see morality as a rebellion against our true natures, instead of as a fulfillment of their potentialities? . . . They [the reductionists] express the integrity and intrinsic dignity of human existence, even as they explicitly disclaim it. Their case is built on a contradiction and cannot prevail (*ibid*, 161,168).

Philip Hefner (1993) offers one answer to Ward's question: the Hobbesian Darwinists who plead for deliberate over-riding of genetic programmes are not yet ready to accept the cultural programmes carried by myths and rituals, because those packets of cultural information defy evolutionary explanation for so long as the existence of cultural group selection is ignored. Colin Grant (1993: 108) points out that Dawkins' plea for a totally unnatural altruism in defiance of the selfish determinations of the natural order is not a repudiation of the modern self-centred perspective but a

> *further instance* of it; because it claims that we are free to assert whatever values we choose over this alien realm of nature.

My view is that Christian theology need have no quarrel with the idea that the evolutionary *roots* of moral behaviour could be linked with self-deception—indeed, it has said much the same, in different words, from the beginning.

Personal righteousness is a noble ideal, seldom achieved, which means that those who claim it are necessarily deluded. The New

Testament writers were quite forthright about that:

> If any think they are religious, and do not bridle their
> tongues but deceive their hearts, their religion is
> worthless (Jam 1:26);
> If we say that we have no sin, we deceive ourselves,
> and the truth is not in us (1 John 1:8).

But evolved morality is not enough. Biological predispositions are real, but they cease at the frontier of the tribe, so that is where Christian theology must take over from cultural evolution. The possibility that the next level of a multi-level hierarchy of selective agencies could produce more co-operative behaviour than the last does not necessarily count as searching for skyhooks (Dennett 1995); on the contrary, to deny the possibility leads in the direction of greedy reductionism. Evolved morality cannot answer the greatest question of all: 'And who is my neighbour?'

4.6 Conclusion: moral history and the environmental crisis

Part of the reason that the environmental crisis is so hard to deal with is that it is so recent, which means that few human cultures have anything in their traditional knowledge that could help us defuse it. This is a dangerous situation, because:

> The novelty of modern social environments is such
> that the proximate behavioural mechanisms which
> were adaptive in pre-industrialised societies are no
> longer adaptive (Irons 1997: 45).

For example, we have deep-seated panic reactions to snakes and spiders, but not to electric sockets or speeding traffic. People now tend to live in much larger groups, in which the dynamics of reciprocity that used to underpin supportive social groups tend to break down. In modern western society it is common to find whole streets full of suburbanites who do not even know who their neighbours are, still less talk to or co-operate with them. Mobility works against co-operation, by undermining the chances that any two people or family groups will be able to establish a long-standing exchange of favours. Secondary and even primary groups break down with increasing frequency, leading to personal disorientation and a pervasive sense of meaninglessness. Traditional myths* are rejected, because they are seen to conflict with other forms of knowledge, especially but not only science, but modern culture offers nothing in their place that has anything like their former vigour and authority. There is no agreed picture of reality as a template for ethics, and no social congruence of ideas about how things are and which things matter—a dangerous state which Rue (1989) calls 'amythia'. EO Wilson's summary of the human dilemma is that we evolved to accept one truth, but discovered another: people, long adapted to believe in gods, have found biology (Wilson 1998).

Most recently, the philosophical/economic ideas of private property and individual freedom have opened up hitherto undreamed-of scales of opportunities for private exploitation, at the same time removing many of the social restraints on personal profit-making. Economic stratification of society reinforces the ancient primate dominance hierarchies, providing powerful incentives to accumulate conspicuous wealth instead of (or as well as) reproductive rights. Self-deception provides ready defences against suggestions (from green activists on all scales from local communities to Rio and Kyoto) that the rich might have an obligation to curb their consumption in order to help the poor. Such defences, going back to Thomas Malthus, suit the efforts of the middle class to feel better than the rest. Charity from the rich, they say, only makes the poor breed all the more, or (the more modern version), the poor are lazy bludgers on the state and don't deserve to be helped. Curbing the lifestyles of millions of people in the rich, powerful economies of US and Australia in order to prevent a few small Pacific island nations from disappearing under the rising seas makes no economic sense.

Only when self-interest is restrained by local interactions and the relentless scrutiny of inescapable close associates can it drive the various forms of co-operation and conditional altruism that underpin the lives of social animals. For humans living in today's anonymous commercial world, those old restraints are weakened. Among the first casualties have been the remaining indigenous peoples of the world, who still try to hold on to the social and religious structures that worked in the past. The ultimate and most significant casualty is, of course, creation itself.

Unfortunately for the environmental movement, it is basically unnatural to humans to think in terms of the global, rather than the local, community (Heinen and Low 1992). We are creatures adapted to small groups, ideally not more than about one hundred and fifty strong. The real danger of our time is that, faced with frightening new threats of unimaginable scope, people will retreat into long-established but now inappropriate reactions (usually traditionally religious or racist). They do no good, but instead drain the moral energy needed for, and inhibit the practical work of, reconciling the interests of individuals, local groups and the global community. If we are to find a solution, we have to do it together, and soon. It will require the sort of mediation that an informed, alert, contemporary church can contribute—a cool-headed but affectionate view of human nature, combined with a theology that encourages people to trust each other and negotiate conservation agreements in good faith. But a mediator has to have the respect of all parties, and that means that the modern churches will have to recover something of the Christian claim to have something rational to contribute to the environmental debate.

5. Theology of Creation

5.1 Introduction

The contemporary environmental debate is having the very good, if unintended, effect of focussing attention on the Christian doctrine of creation. That attention is often very critical, because the Christian attitude to the natural world is widely perceived to be exploitative and inseparable from that of the pre-scientific age in which it was formulated. Yet, as Peacocke (1979: 46) points out,

> any affirmation about God's relation to the world, any doctrine of creation, if it is not to become vacuous and sterile, must be about the relation of God to, the creation by God of, the world which the natural sciences describe.

Christians quoting Biblical authority therefore often encounter a credibility gap when making public statements on environmental matters, and are not likely to have much influence in the secular debate until this problem is resolved.

The inevitable starting point for any re-examination is, of course, the two creation stories of Genesis, and other relevant texts on creation theology in the Old Testament. Psalms 19, 24, and 104, and Job 38-41 contain important general statements about the Biblical attitude to nature, and the Pentateuch includes some startlingly modern-sounding specific commands. For example: Deut 20:19 prohibits deforestation as a military tactic, asking 'are the trees in the field men that they should be besieged by you?' According to Ex 23:12, the purpose of the Sabbath is partly also to allow rest days for working animals. Psalm 144 denies human hubris with 'man is like a breath, his days are like a passing shadow'.

The Biblical world-view is very different from ours, but we must treat its message with serious, critical intelligence. As Hall (1986: 174) points out,

> We can hardly expect the Biblical authors to have
> spoken directly to the crises we confront. They did not
> have to deal with a beaten natural world, or with
> human beings whose technical genius had brought
> them to the point where they can destroy trees not only
> with axes, one at a time, but with acid rain, from afar,
> and by the millions! The remarkable thing is that there
> is as much as there is in this ancient literature by way
> of compassion for nature and the recognition of human
> solidarity with it.

Hall adds that one of many problems associated with using the
Biblical evidence is that it ' . . . already participates in the 'ambiguity'
[about nature] that has typified the history of Christendom' (as
comprehensively documented by Santmire 1985). But beneath the
multiple witness and the historical context of the Bible there are deeper
strata of Biblical faith, and the recovery of them is one of the primary
tasks of critical and constructive theology in modern times.

5.2 The classic Christian doctrine of creation

All images and ideas about creation are necessarily shaped by the
contemporary knowledge and experience of those who formulated
them. The authors of the Biblical creation stories, the ancient Hebrews,
worked in a cultural milieu dominated by the mythologies of Egypt,
Canaan, Assyria, Babylon and Persia. Through interactions between
these traditions, going back to the fourth or third millenia BCE, the
Hebrews developed their fundamental conviction that creation is to be
understood in terms of

> the unique sovereignty of Yahweh, the God of Israel,
> and the complete subservience of all nature, both in
> heaven and earth, to . . . a single code of law which was

126

established along with the universe at the beginning of
time (Kaiser 1991: 6-7).

The Hebrews' faith in Yahweh ('the God who creates continuously')
was forged out of their belief, largely influenced by their experience of
the Exodus, that the God of Abraham was stronger than the gods of
their neighbours. During the exile, they developed a wider
understanding of the whole world as radically dependent on a constant
upholding by divine action (Anon 1994: 6). They believed that the only
reason that the universe exists is that God delights in it for itself and
positively wants it to exist—and if God were ever to cease upholding it,
there would be nothing. As Cupitt (1984: 57) put it,

> the idea of Nature as a whole, which we owe to the
> Greeks, they [the ancient Hebrews] simply lacked.
> Where we see a natural world they saw the effects of
> the activity of God . . . all things revealed God, as
> drapery reveals the movement and activity of the body
> inside it.

For the Hebrews all discussions of creation concerned a practical
and ontological, not a temporal, question, about how things are, not
about where they came from. To the Hebrews,

> the origin of the universe was beyond human
> understanding . . . but its subsequent operation can be
> understood due to the fact that human reason is in
> some way a reflection or image of that same lawfulness
> or reason that governs the world (Kaiser 1991: 6).

The historic (Biblical) creationist tradition is therefore far older and
more fundamental than the recent concept of 'creationism' espoused by
modern fundamentalist sects, and the two have very little in common.
Kaiser (1991; 1993) summarises the three main themes of the historic
tradition as:

1) the natural world is orderly, comprehensible and
accessible to human understanding;
2) contrary to previous ideas, heaven and earth are
made of the same sorts of matter;
3) nature is relatively autonomous, operating
according to self-sufficient laws.

It is of course essential to be careful when making comparisons over
vast spans of time. Not even Biblical doctrines behave as constants,
unchanged over thousands of years. Substantial shifts in the primary
emphases of Old Testament creation theology were introduced over later
centuries.

First, the New Testament writers worked in a totally different setting
from that of the Hebrews, dominated by Hellenistic, Egyptian, Syrian
and Iranian cultures. Much of the other-worldly attitude that has shaped
the Christian response to nature is an import from Greek philosophy in
New Testament times (Santmire 1985); hence, the New Testament
scriptures are much less earthy than the Hebrew.

Second, the idea of *creatio ex nihilo*, creation out of nothing, did not
appear until Maccabean times (Simkins 1994: 178), about 200 years BCE,
and was not incorporated into church doctrine until the end of the
second century CE (McGrath 1994: 234). None of the great prophets and
wisdom writers of the Old Testament were aware of it, although it
became very influential in the Hellenistic world of the New Testament.
It so happens that the Priestly creation story in Genesis 1 is broadly
compatible with contemporary scientific cosmology, which sees the
origin of the universe in the so-called 'Big Bang', but that is not the point
of P's story. Neither does such a simplistic interpretation do justice to the
range of models of the relationships between the human and natural
worlds employed by Biblical writers in different times and historical
circumstances (Simkins 1994: 255).

Unfortunately, during the course of history the balance between the
two interlocking aspects of creation theology, concerning origins versus
dependence, became lost. By at least the end of the eighteenth century
(Berry 1995: 21), and certainly by Darwin's time, Christian creation
theology was emphasising the idea of an original act, the *beginning* of

life, much more than that of a continuing process of upholding the *ongoing conditions* for life. That distortion of the Biblical insight caused great problems for Darwin, and eventually it helped to undermine his faith, as it still does that of many young people brought up in modern Christian homes.

The difference between the concepts of dependence and origins in creation can be illustrated by analogy with a television drama. What viewers see is the joint work of the author of the script, the director, the actors, the production crew, and the developers and manufacturers of the transmitting and receiving technology. The origin of the play can be analysed in terms of all their different contributions, and that would certainly be a complex enough task in itself. But ultimately, all these participants in the chain of events, as well as the viewers themselves, are dependent on a steady supply of electricity that is taken for granted and yet can be interrupted at any moment—and when that happens, everything stops. The electricity does not itself directly create the play, any more than God directly creates blackbirds probing a lawn for worms, but it constantly upholds the conditions under which all the complex work of organising the writing, performing and enjoyment of television plays can be done. Independent human actors and playwrights can do their creative work well only when they are free to assume that the power that brings their productions to life will always be there, as God constantly upholds the conditions permitting all the multitudinous and unconscious activities of blackbirds, grass and worms.

5.3 Creation theology and the origin of modern science

Unlike most other cultures of their time, the Hebrews insisted that trees, rivers and rocks did not have their own resident spirits, but that all matter was merely matter, open to human use and investigation. Christian creation theology inherited this attitude, and is therefore seen to have been responsible for a systematic, historic campaign to demythologise nature. White (1967) severely criticised Christianity for this doctrine, on the grounds that it removed the protection that superstition had once afforded the natural world, and opened the way to the unrestrained exploitation that has produced the modern ecological crisis. Yet that very same demythologising doctrine also laid the foundations of modern science (Turner 1998).

The scientific view of the rational, ordered universe is entirely compatible with the theistic, Christian affirmation that we can make sense of the world because God's faithfulness stands behind it. The three main themes of the historic creationist tradition asserted that the universe reflects the goodness, rationality and freedom of God and therefore creation itself must be good, rational and contingent. In due course these fundamental attitudes were incorporated within Christian faith.

Christianity was therefore open to science from the beginning, and this indeed is one of several reasons why the roots of modern empirical science are deepest in the Protestant Christian west (Wolpert 1992: 46; Barbour 1997: 28;Turner 1998). Other cultures had developed systems of organised knowledge over many centuries, but no ancient science produced western-style, experiment-based technology. The Greeks were hampered by the Platonist view of all matter as inferior embodiments of pure rational forms, so they believed that experimenting on them would be pointless. The Chinese had no geometry, and their holistic view of life precluded the development of the reductionist analysis required to develop technology (Wilson 1998). Religious authorities in Roman Catholic and Islamic countries, and state officials in China, exerted tight control over higher education (Barbour 1997: 27); in contrast to the intellectual freedom of western universities, especially those in Protestant countries.

For Newton and many other early members of the Royal Society in London, science was a means of reaching out to God. The three theological concepts of goodness, rationality and contingency are among the vital foundations of science. If the universe is good, it is worthy of careful study; if it is rational, it is predictable and reliable; and if it is contingent it could have been otherwise than it is, so the state of things has to be studied by experiment, not deduced from pure reasoning. Moreover, the Hebrews insisted that there had to be a fruitful balance between the rationality and the freedom of God in creation: if rationality is overemphasised, the universe becomes fixed and uninteresting, whereas if freedom is overemphasised, the universe becomes incoherent, unpredictable and impossible to study.

In a nutshell, if the world is not rational, science is not possible; if the world is not contingent, science is not necessary (Newbigin 1986: 70). The Christian theology of creation claimed that it is both. Thus the historical relationship between theology and science in the western world has been very much more long-standing, complex, productive and positive than many participants in the present debate may realise. On the other hand, Christianity should not, and does not need to, defend itself by claiming credit for having contributed to the rise of science, which would expose it to the developing contemporary backlash against the excesses of scientific technology. The most it need claim is that true Christianity is not, and never has been, incompatible with true science (Peacocke 1979: 76; Ward 1996).

C S Lewis neatly illustrated this compatibility when he put into Screwtape's mouth the advice (to a young devil attempting to ensnare an unsuspecting human soul),

> Above all, do not attempt to use science (I mean, the real sciences) as a defence against Christianity. They will positively encourage him to think about realities he can't touch and see. There have been sad cases among the modern physicists. If he must dabble in science, keep him on economics and sociology (Lewis 1942: 14).

Sound advice indeed—and in view of the diabolical consequences of modern market economics, one might deduce that Screwtape's pupil has been remarkably successful in following it.

Since Biblical times, according to Moltmann (1985: 33-34), the relationship between the doctrine of creation and science has passed through three distinct stages.

1. During the first stage, Biblical traditions and the ancient world's picture of the universe were fused into a religious cosmology emphasising a divinely ordered world filled with God's glory and guided only by divine wisdom. The consequences of this doctrine for

the attitudes of ordinary people to the animals and plants around them have been well summarised by Thomas (1983). This stage has no modern equivalent except among extreme literalists.

2. The second stage began when the sciences emancipated themselves from this cosmology, while theology detached its doctrine of creation from cosmology and reduced it to a personal belief in God as Creator rather than in the things that have been created. The two disciplines established, after many struggles, their own identities on either side of accepted demarcation lines, and achieved a peaceful co-existence based on mutual irrelevance. This stage is roughly equivalent to that of Barbour's Independence model (p.19).

3. In the third stage, which we are just entering now, the sciences and theology are becoming companions in tribulation, under the pressure of the environmental crisis and the search for new directions in both which must be found if humanity is to survive at all. In this newly co-operative atmosphere, the mutual demarcation lines are no longer necessary. In a global situation where it is one world or none, says Moltmann, science and theology cannot afford to divide up the one, single reality. His view is confirmed by the success of the Templeton Foundation in stimulating intellectually rigorous courses in science and religion in tertiary institutions around the world.

Many academics and students now see the two disciplines as partners to be taken seriously. This stage includes Barbour's Dialogue and Integration models (pp.20, 21), and it brings great hope for the future. Allied with science, there is every reason to hope that Christianity can make a useful contribution to the environmental debate. Science emphasises the dynamic aspect of creation which theology had temporarily forgotten, and at the same time is raising all sorts of questions which are outside its own province to answer. For example, modern medical science encounters many life-or-death dilemmas in which science and ethics cannot avoid meeting: and the solutions are often rooted in the Judaeo-Christian tradition. Contemporary developments in creation theology are helping to cast intellectually respectable and ethically relevant light on these dilemmas.

5.4 What is 'the integrity of creation'?

The concept of the 'integrity of creation' introduced at the 1983 Vancouver Assembly of the World Council of Churches was new and not well defined, but generally agreed as having implications far beyond the earlier idea of sustainability (Gerle 1995: 47). To some, its very novelty was an advantage: one participant in a discussion of what it might mean commented, with some exaggeration, 'Thank God the ecumenical movement has finally given us a concept without content so that we can put into it what we want to!' (Niles 1989: 58). To others, the concept of the integrity of creation is too indefinite to be at all useful— the German delegates at Vancouver declared that it is untranslatable, and the Basel conference avoided the term altogether (Gosling 1992: 9). A scientist wants to know what a theologian understands by it, and further, what 'safeguarding' it means—from whom or what does it need safeguarding, and why?

The original meaning of integrity, as understood in the Hebrew scriptures, was primarily a matter of relationship to God and adherence to God's laws. The integrity of nature consisted in the fact that it had never disobeyed God; nature was always and still is perfectly obedient to God, so if any part of nature was hostile to humanity after the Fall, that was because it was being used as a means of conveying divine judgement on humans (Kaiser 1996). If we have lost this sense of the integrity of creation in modern times, or act as though it did not exist, that is *our* problem, says Kaiser: a problem with our concept of creation, and underlying that, a problem with our concept of God. He agrees with Pagels (1988) that the later Christian ideas, that the fall of Adam necessarily involved the corruption of the whole natural world and therefore that Creation needs redemption as much as humanity, are simply wrong and are to be discarded.

It is a surprise and a relief for a biologist to be assured by theologians that the Hebrews understood that thorns and thistles are an integral part of God's creation, and that it is only their *presence* in tilled fields that made them a curse to the Hebrew farmers, not their own characters; that only arable land was cursed, not all of wild nature; that if there was a conflict between God and Israel, nature was on God's side; and that the groaning of the earth whilst waiting for the sons of God to appear was less about loss of integrity than a poetic image concerning the corruption of bodies in their graves. These concepts translate much more readily to the modern mind than the fantastical Greek ideas that came to overlie them in New Testament times. Kaiser completes his reinstatement of the original and far more contemporary-sounding and satisfactory meaning of the phrase by putting humans in their place:

> The integrity of nature . . . is not something humans can either contribute to or detract from . . . if we fail to respect [it], we will certainly bring ruin upon ourselves . . . but the order of nature—as understood in Scripture . . . owes no more to human participation than it does to any other species (*ibid*, 290).

In other words, the Biblical view is that creation needs safeguarding from *humanity*, because of the two, only creation retains its original integrity intact.

Within WCC, the idea of the integrity of creation has proved useful, for two reasons summarised by (Niles 1989: 58). 'First, *it has given a new prominence to the doctrine of creation'*, understood as the rather difficult theistic* idea of a continuous and continuing divine upholding of all life, rather than simply as the deistic* idea of 'the initial divine act which set nature and history on their course' (p.127). Second, since the Biblical idea of the integrity of creation refers to much more than ecological issues, it offers a context for our struggles for peace and justice. Hence, the new term 'integrity of creation' is widely understood to mean that there is a 'moral order given in creation that we disregard or violate at our own peril . . . moral value or worth [is] something the Creator has bestowed on the whole of creation[1] and not just on the human part of it . . .' (Niles 1989: 59). The natural inclusiveness of the concept that creation has its own integrity provides a theological means of holding together the issues of justice, peace and environmental management.

At Annecy, in France, a meeting on theocentric ethics in September

---

1. This concept is treated at much greater length by Murphy and Ellis (1996).

1988 (Birch and others 1990: 277) agreed, after extensive discussion, on a working definition:

> The value of all creatures in and of themselves, for one another, and for God, and their interconnectedness in a diverse whole that has unique value for God, together constitute the integrity of creation.

The emphasis of this definition appears to be upon creatures as individuals, which is understandable from a theological point of view, but it is likely to run into difficulties when conflicts arise between the good of the individual and the good of a larger entity, such as the ecosystem. For example, what happens if deer, seen as 'creatures of value in and of themselves', become so abundant that they cause damage to their habitat, which presumably is included within the idea of 'a diverse whole that has unique value for God'? In such a situation, which is common in parts of Scotland, New Zealand and USA, the two sets of values are incompatible. Which of them is to be sacrificed for the other?

For Aldo Leopold, widely regarded as the father of wildlife conservation in USA, the answer is clear. He was a hunter and woodsman of great skill, and in *A Sand County Almanac* (Leopold 1949), he advocated the view that, when necessary, the individual must be sacrificed for the ecosystem.

> All ethics so far evolved rest upon a single premise: that the individual is a member of a community of interdependent parts... The land ethic simply enlarges the boundaries of the community to include soils, waters, plants, and animals, or collectively: the land (203-04).

> A land ethic of course cannot prevent the alteration, management, and use of these 'resources', but it does affirm their right to continued existence, and, at least in spots, their continued existence in a natural state. In

short, a land ethic changes the role of *Homo sapiens* from conqueror of the land-community to plain member and citizen of it. It implies respect for his fellow-members, and also respect for the community as such (204).

A thing is right when it tends to preserve the integrity, stability, and beauty of the biotic community. It is wrong when it tends otherwise (224-5).

What Christians need is a concept of the integrity of creation that transcends the static and simplistic theological ideas of 'wholeness', or even 'telos' [purpose]: one that does justice both to (a) the restless dynamism of the real natural world recognised by science, and to (b) the real conflicts of interests between individual and community at all levels of nature.

There is little recognition of that need so far. For example, in early 1988 a large meeting of the WCC on the theme of JPIC was held at Granvollen, Norway. Granberg-Michaelson (1994: 99) commented on the extent of the agreement among participants on the central ideas, including that 'Every creature and the whole creation in chorus bear witness to the glorious unity and harmony with which creation is endowed'. Such a perspective would sabotage any prospect of co-operation with evolutionary biologists on environmental issues. Likewise, theologians may expound the ethical idea of encouraging the 'flourishing' of natural systems as a theological aim. But in places like California or New Zealand, both much altered by the flourishing of introduced species, this perspective offers no precise answers to the question of how we should distinguish between flourishing pests, to be controlled, and flourishing native ecosystems, to be encouraged.

It is important and encouraging to note that Leopold's secular definition of the integrity of 'the biotic community', Annecy's theological idea of the value of 'a diverse whole that has unique value for God', and JPIC's term 'the integrity of creation' all avoid the anthropocentric term 'environment'—always taken to mean the environment of humans. The creation that we are concerned to care for, or at least to avoid damaging, has its own life and values, and is far more extensive than the human environment—it existed long before we evolved, and in vast areas of ocean, mountain ranges, deserts and tundra it still survives independently of us. Recognition of this is a real contribution to contemporary culture (Gerle 1995: 117). Nevertheless, secular literature (ie, most literature) is primarily concerned with the human environment, and, as it provides an important resource for this study and its focus cannot be changed merely by inserting a different terminology, I have continued to use the term as generally understood.

Gosling (1992: 10) points out that 'integrity' is a *relational* word, with both vertical and horizontal dimensions. Theologically,

> integrity of creation implies both the vertical dependence of creation on its Creator and the worth and dignity of creation in its own right (i.e. its intrinsic value) . . . [horizontally], every creature is bound to every other creature in a community and communion of being. Human beings especially must recognise that we are not separate from and above the rest of creation, but part of its totality, sharing with other living beings their limitations and destiny.

Such a concept is already there in Leopold's definition, but is new to mainstream Christianity, and if accepted would entail a huge revision of many ancient theological convictions concerning the dignity of humanity. Even in the secular world, it would meet a great deal of resistance.

5.5 Current developments in creation theology

If the secular theory of gene-culture co-evolution is valid (p.80), it should be able to cast new light upon old problems that have previously been analysed only in different terms. The Christian message has been studied for centuries, yet it still raises questions that have puzzled generations of philosophers.

Some of these questions are relevant to the environmental crisis, and cannot be explained by assuming that humans are rational or can learn to be altruistic. For example, 'Why is it such hard work expanding the bounds of affection—loving one's neighbours and enemies?' (Midgley 1978: 344). Or, responding positively to the Seoul affirmations (p.32)? We have the ideal opportunity to concentrate on such questions now, because any public promotion of the the the idea of religious input into the environmental debate will inevitably invite inspection of Christian ideas about creation, and it would be as well to check on our contemporary understanding of it first.

Like all religious models, the doctrine of creation is neither a literal description nor a useful fiction, but a human construct that allows us to interpret our experience by imagining what cannot be observed (Barbour, 1997: 119). The critical realist approach (p.25) requires us to take this and all theological models seriously but not literally; not as a static entity, but as a continually adjustable, fruitful source of improved understanding of pre-existent reality. How much progress have the churches made in this challenging but necessary process?

If there ever was a time when it was possible to think about theological questions with the calm civility associated with venerable institutions and ageless traditions, that time is now long gone. The challenge of various new paradigms, especially feminism, to all forms of academic work has already shaken the ancient traditions. Now the additional, even more urgent environmental crisis that faces all humanity has simply removed the easy option of regarding the future as a predictable extrapolation of the past. It is too late to study systematic theology hundreds of years in the making (as taught in traditional theological colleges) and ask what it can say about the contemporary problems that did not exist, or were not recognised, until twenty or thirty years ago: now we are being forced to look at contemporary problems and ask what they mean for the reform of systematic theology.

That reform is beginning, and, so far as it goes, it is immensely exciting—the Bibliography cited here gives a flavour of it—but it is achingly slow. Debate in theological colleges on the feminist critique of ancient doctrines is much more intense and advanced than debate on the environment crisis. Theologians should be working hard to inform policy-makers and public of the spiritual and social consequences of current demographic and ecological forecasts, but, like the scientists criticised by

Ehrlich (1997), too many devote themselves to ever-more sophisticated analyses of less urgent questions. At parish level, environmental matters are usually regarded as a side issue, and get rather little attention in comparison with the much-higher profile arguments surrounding internal church politics.

In the contemporary world, global conservation is a multi-cultural issue. A Christian theology that is to avoid sabotaging dialogue with people of other faiths or none must shift into a new gear, since 'there can be no responsible theology now that is not global in its perspective', says Hall (1986:41). On the other hand, the convictions that motivate any particular party to the debate are their own, and the better understood they are, the more likely that party's contribution is to be convincing and helpful. It is therefore vital for Christians involved in the debate to appreciate the unique perspective brought by Christian theology to the problems of global conservation.

All credible theologies must take into account the view of reality current in their own day (McFague 1993: 73-4; Peacocke 1993: 7). Over the long history of Christianity, that has meant periodic reformulations of doctrine: so Augustine integrated Christian concepts with the world view of neo-Platonism, Aquinas with that of Aristotle, and Paley with that of Newton (Kaiser 1991). In any age, when the secular picture of reality undergoes a significant paradigm shift, theology must attend to it, despite its general suspicion of innovations (Küng 1989). It may take a while: the Roman Catholic Church officially reinstated Galileo only in 1992, and recognised the Darwinian theory of evolution as valid four years later (John Paul II 1996).

The challenge has already been taken up by, for example, the feminist and process theologies of today. Not everyone will agree with their positions, but they have helped the churches recognise that science and theology *both* depend on mental constructions and metaphors to create models of invisible reality that are *in principle* open to correction, and that is making possible a new phase of serious discussion between scientists and theologians.[2] The introduction of Darwinian perspectives

---

2.   Not all members of either camp are included, of course: but fundamentalist Christians on the one hand, and hard-line reductionist scientists on the other hand, represent the two ends of a broad spectrum of views that includes very many open-minded enquirers not committed to either extreme.

into traditional creation theology may well be resisted in some quarters, more so at parish level than in theological colleges, but it does not require nearly so drastic a revision of old ideas as has feminism, which has already demanded reinterpretation of the Eden story for reasons of its own (Pagels 1988). Some authors call for a restructuring of the entire framework of Christian thought about the relationships between God and the world (Gosling 1992: 49); others maintain that the only change required is a more faithful understanding and preaching of established doctrines (Berry 1995: 40). The most recent, thorough and authoritative treatment opts for 'starting all over again' (McGrath 1998), because

> what once seemed as if it might be a wonderfully creative and interesting discussion appears to have degenerated into little more than a slanging match between a group of natural scientists bent on eliminating religion from cultural and academic life, and a group of religious people who seem to know (and care) nothing for the natural sciences. What the Rennaissance envisaged as a dialogue has degenerated into what is depressingly often a mutual display of ignorance, hostility and spite (7).

The World Council of Churches is at the forefront of the effort to redefine the relationship between humanity and creation. A recent summary of its activities (Chial 1996) placed the 'need to re-articulate theology of creation' as central to this concern. However, Chial laments that no coherent new formulation has yet emerged, nothing that inspires change; the prevailing version

> remains bound to western industrial and economic models that seek to manage and control . . . the ecumenical movement has barely begun to . . . provide an ethical approach to the tensions between ecological concerns and social needs. Nor have churches in the North had much impact in the area of curbing the consumerist life-styles that remain at the heart of their

governments' opposition to global environmental and
economic agreements (53-4).

In my view, no ecumenical reformulation will make much headway
until it takes seriously what biology can say about the influences of
game theory and of the unconscious roots of human morality on
individual ethical decisions (pp.71-123, 163).

The Christian theology of creation asserts that the world is:

intelligible,   because its fundamental laws were set by a supremely
rational creator;

reliable,   because the faithfulness of God stands behind it and
continually upholds it;

sanctified,   by the continual presence of God in it, and by the
incarnation in historic time;

indispensable, because it is our meeting place with God in contemporary
time;

sacramental,   because it is the outward and visible sign of God's love
and energy; and

significant,   because the freedom of all creatures in the present is real,
and their use of it determines what will be available for
God to take up into the new creation at the end of time
(Page 1996).

These beliefs help to develop very positive attitudes to life in general
and to environmental matters in particular. People who share them will
feel confident and 'at home in the universe', recognising but able to face
the modern and psychologically devastating uncertainties projected by
extreme doom-sayers and nihilists. They will think of all creation as in
some sense hallowed by the presence of God, so will be unlikely to
abuse it or treat it as having merely instrumental value, and will not
give up on caring for it or on opposing the forces of destruction. The
world makes ultimate sense to them, despite its many and very real
agonies, because it is not a self-contained or meaningless system but has
behind it the purposes and the faithfulness of God.

The literature describing modern developments in creation theology
is recent but huge, and much of it is well summarized in recent texts
(Barbour, 1997; Peacocke, 1993). It includes extensive, sympathetic but

critical re-examination and updating of the formulations of ancient wisdom in the light of the Darwinian understanding of nature. This immense task is now being actively pursued by contemporary theologians of most Christian denominations as the evolutionary and physical sciences progress (eg Hefner 1993; Page 1996; Peacocke 1993; Ward 1996). Their work encourages rational faith in the idea that, *provided* the churches begin actively to teach the Biblical doctrine of the integrity of creation in its authentic form (p.132), and to correct the centuries of misinterpretation denying the compatibility of Christian creation theology with science (p.129), it has the potential to provide the macrophase* wisdom that our society so urgently needs. On the other hand, we cannot assume that merely providing that information will be enough. Forty years ago, Teilhard de Chardin proposed extensive doctrinal reformulations in the light of the evolutionary biology of his time, but with rather little impact upon mainstream theology.

Our models of the general characteristics of nature will affect our understanding of God's relationship to nature, and hence our interpretation of creation and redemption—the enduring themes of Christianity.

> Nature today is understood to be a dynamic evolutionary process with a long history of emergent novelty, characterised throughout by chance and law. The natural order is ecological, interdependent and multileveled. These characteristics will modify our representation of the relation of both God and humanity to non-human nature. This will, in turn, affect our attitudes to nature and will have practical implications for environmental ethics. The problem of evil will also be viewed differently in an evolutionary rather than a static world (Barbour 1997: 101).

Churches speaking to informed audiences can no longer hold as rational any concept of creation that involves a precise, hands-on, active moulding of living creatures by God. Jeremiah's prophetic image of the potter (Jer 18:1-12) shaping and reshaping a vessel is commonly misinterpreted by fundamentalists as a picture of God working on Creation. Matter is seen as totally inert, shaped only by the hands of the potter, owing nothing to any random process and incapable of any initiative of its own. In fact, Jeremiah's point was a totally different one, calling the House of Israel to repentance in order to escape impending disaster, and warning that God could destroy Israel and start again, just as a spoiled pot can be replaced by a good one. Theological doctrines always start as human reflections, so Jeremiah's interpretation of the Old Testament view of God's relationship to creation, still accepted by literalists, conditioned his view of the potter. Equally, Darwin's interpretation of ruthless nineteenth-century industrial competition probably conditioned his theory of natural selection (Grant 1993).

The current view is that living beings are neither inert nor the product of solely random processes. Rather, it is the *interaction of chance and law in an orderly universe* which provides the crucial conditions for natural selection to operate. Peacocke (1979: 95) describes chance as 'the search radar of God', sweeping through all potential fields of statistical possibility. Without chance there could be no new variations; without order, all new variations would be immediately lost. Favourable variations are accumulated over many generations, and evolution—changes in gene frequencies from one generation to the next—is the result (Appendix 1). It is the *consequences* of natural selection that produce new life forms.

Models of natural selection based on population genetics concentrate on the inheritance of information (the codical domain), but theology is concerned with the material domain, the scene of all our lives, experiences and morally significant decisions (p.194). In my view, theology need have no qualms about accepting the concept of cumulative change by natural selection. It is not, as Williams (1996: 156) asserts, an 'evil . . . unreasoning enemy',[3] but a statistical, often very small, difference in the probability of reproductive success between slightly different individuals. The real freedom that God has given to all life places no limits on the actions of natural selection in the natural world; nor, presumably, on the multi-level processes which led to the

---

3. Darwin himself called natural selection 'clumsy, wasteful, blundering, low and horribly cruel' (quoted by Rolston III 1994).

emergence of cultural selection in the human world, and eventually to the abandonment of selection in the spiritual world (p.150). Moreover, I would assert that God accepts the consequences of creaturely freedom, and loves the resulting riotous variety of amoral natural life presumably as much as God loves the variety of moral and immoral human life that evolved from it. God created, not individual organisms, but the fundamental laws by which, first the universe itself, and then living organisms within it, could create themselves.

This is not such a modern idea as it sounds: it was already foreshadowed in the historic creationist tradition (p.126). Augustine himself clearly recognised that the creation brought into being by God was gifted from the outset with the capacity to assume a rich diversity of forms (van Till 1996). Although Augustine was not prepared by his intellectual heritage or environment to consider the idea of evolution as we now understand it, his picture of a creation gifted with self-actualising potentiality is certainly not incompatible with Darwinism. Indeed, I venture to guess that Augustine would have been appalled by the modern 'scientific creationists'. Like all his contemporaries he understood Genesis 1-3 literally, but he also insisted that

> the literal meaning . . . may never stand in contradiction to one's competently derived knowledge about the earth, the heavens and the other elements of this world . . . [that are] certain from reason and experience . . . Augustine soundly reprimands those Christians who defend interpretations of Scripture that any scientifically knowledgeable non-Christian would recognise as nonsense (van Till 1996: 30).

Page (1996) has put that ancient insight into the context of evolution, arguing that what God created was *possibility*, without strings or conditions of any sort—she does not even allow that God might have nudged evolution along in a preferred direction, such as that leading to humans—and the world we see is the response of living things to the possibilities opened to them. God's action in the world is defined as '*a continuous giving room to explore what is possible*' (*ibid*, 17), not a direct action of making any particular thing . . . So 'the sense of wonder remains, but is redirected from a "something" to the whole changing variety of what has been and now is in the world' (*ibid*, 4). Chance is part of the design, not incompatible with it. The freedom it provides makes love possible, she says, and is the indispensable prelude to the relationships between God and creation which grew from this freedom. God is passionately involved in those relationships at every level.

The trouble is, freedom produces

> human and non-human suffering from the way in which change and variety in creation have unpredictable and unfortunate results . . . Because God is involved in the relationships, the pain as well as the pleasure is part of the divine experience. Yet it cannot be relieved by God, since that would remove the fundamental freedom to use possibility, which is the gift of creation (Page 1996: 105).

On this view, it is merely a logical consequence of creaturely freedom that, when a lot of independent and non-sociable animals pursue their own self-interest at the same time and place, many forms of evil are the direct result. A modern creation theology that has recovered Irenaeus's ancient perception of creatures as necessarily imperfect (Brown 1975) sees evil as the unavoidable consequence of freedom: God could not have prevented it (Hick 1977). This is the simple and obvious answer to Gould's concerns about the non-moral nature of ichneumon wasps (Gould 1983b). The problem of evil in nature, for the biologically-informed theist, is not the existence of evil, such as the apparently useless suffering of the ichneumon's caterpillar victims or of the last-hatched 'back-up chicks' of the pelican (McDaniel 1989), but the *extent* of it. Contrariwise, for the reductionist sociobiologists, the most puzzling problem is the existence of genuine goodness.

The Christian doctrines of incarnation and redemption emphasise the Christian belief that the material creation *matters*.[4] Evil, or at least, disvalue, may arise from matter (Rolston III 1992), but (in clear distinction from many rival theologies of New Testament times) matter is *not* itself intrinsically evil; the world is our home and is sanctified by the presence of God in it.

The original Hebrew concept of creation did not include the idea that nature was also corrupted by the fall of Adam (Kaiser 1996). That idea, although applied only to to arable land and human nature rather than to the whole of wild nature (Kaiser 1996), was added by Augustine, via a train of logic that seems bizarre to us. Modern philosophers agree that, regardless of Augustine, no part of the cosmos needs redemption in the same sense that humanity does. Nature's problem is not the guilt of human sin, but the consequences of it (Rolston III 1994). On the other hand, the cosmos would certainly benefit from human redemption to the extent that redeemed humans learn to care better for creation. Our material life and environment, once shared by the Word of God, are thereby affirmed as worthy and to be cherished; and a cosmos loved enough by God to be thought worth visiting in person deserves our greatest care. Teilhard too argued that sin and evil are both inevitable consequences of the slow creative processes of evolution, which means that creation and redemption are both included in a single process (Barbour 1997: 248).

Physicists are in no doubt that at some point in the relatively near future (about ten million years, according to James Lovelock[5]—that is, less than a quarter of the time since the extinction of the dinosaurs and the rise of the mammals)—the unstoppable rise in global temperature will extinguish all life. The last survivors are more likely to be thermophilic micro-organisms than complex and vulnerable mammals such as humans. But that does not mean that conservation is pointless: as Leakey (1996: 253) argues, ' . . . the fact that one day *Homo sapiens* will have disappeared from the face of the Earth does not give us licence to do whatever we choose while we are here'.

The contemporary process of rethinking of Christian attitudes to the natural world has to be done in the context of the other experiences of and knowledge about the world which people have to integrate with

---

4. It is an ironic contrast, says Ward (1992: 146) that materialism 'takes much too low a view of matter'.
5. In response to a question following a lecture given at Green College, Oxford, February 1997.

their understanding of theological teachings, and that means greater integration of theological and secular expertise. For example, arguing against traditional belief in, say, the Garden of Eden or the Fall is logically necessary in modern culture, but does not destroy the ancient insight that humans are naturally sinful. Adam's sin is no longer needed to explain human failings, since sociobiology explains them better (Williams 1998), but shifting to that explanation requires establishing a link with evolutionary science. That is no problem for a scientist, but it is a real stumbling block for those whose knowledge of biology and earth history is hazy. We need to find ways of doing that in terms that mean something to ordinary people. The thought of involving the laity might be scary to professional theologians, but, as Schmitz-Moormann (1995) points out, tongue firmly in cheek, it should be no more impossible for theology to involve non-theologians than for God to raise sons for Abraham from stones. After all, in so far as theology has access to raw data comparable with those of the sciences, those data consist of the religious experiences of ordinary people (Murphy 1990; Peacocke 1995).

Moreover, theological reforms are less likely to be successful if they do not carry the people with them: as (Cupitt 1984: 25) points out

> In England . . . religious change always began with the King and the Court, who then simply imposed the new order on the country . . . as a result the English have a long tradition of resenting . . . a religious order imposed upon them from above.

New Zealanders share some of this attitude, as was well demonstrated during the negotiations over the 1971 Plan for Union.

Much of the resistance encountered in applying scientific ideas to human affairs comes from prejudice against science, almost always due to specific misconceptions. An unpublished American survey found that an astonishing forty-five per cent of twelve hundred first-year university students rejected evolutionary science (Zacks 1997). Their objections were not primarily due to religion, but to demonstrably false interpretations of standard theory, such as 'mutations are never beneficial to animals'; 'the methods used to date fossils and rocks are unreliable'; and 'the chance origin of life is a statistical impossibility'. Some unknown number of New Zealand students may have similar problems.

Poor science education among congregations is relevant here, because Christians need to understand some basic science if they are to take the environmental crisis seriously. People who still believe that God set limits to the sea[6] and made the hills to stand eternally are not well equipped to deal with the problems arising from global warming, rising sea levels and massive soil erosion.

Those who are already well versed in science may have a different set of problems with integrating evolution and religion. Prejudice arising from respect for well-known authorities critical of religion such as Richard Dawkins will certainly make promotion of church involvement in the debate difficult, just as Lynn White's critique of Christianity influenced the previous generation of students, including Max Oelschlaeger (1994). Dawkins' metaphor of the selfish gene is widely known but misleading (p.101), and feeds the conflict model by seeming to present religion and science as *alternatives*. In fact, the opposite to religion is *materialism*, not science itself (Ward 1996).

By contrast, the idea of emergent properties and multi-level selection, which allows for true Christian ethics to evolve from gene-centred processes (p.112) is not yet widely *enough* known. In either case, it seems that church-sponsored education programmes in creation theology will have to start with basic evolutionary science. Perhaps the education process should go both ways: as Holmes Rolston (1994) remarks, perhaps theologians need to figure out what they believe before they talk to biologists; perhaps theologians will not be able to figure out what they believe until after they have studied biology.

---

6. This is not only a matter of education in science but also in theology. To the Biblical writers, the sea represented chaos, which is why, in the book of Revelation's vision of the new earth, 'the sea was no more' (Rev 21:1).
   We should not deduce from this that heaven is no place for those who love the sea and its creatures.

5.6 The origins of true Christian ethics

Secular evolutionary theorists such as Alexander, EO Wilson and the reductionist school propose that emotions such as love, loyalty, sympathy and gratitude are also products of natural selection, along with legs and wings (p.109). Their gene-centred view of natural selection has some interesting implications for any updated version of Christian theology. If our emotions are the conscious feelings that lead us to make decisions consistent with the biological processes (kin selection and reciprocity) that favour long-term *genetic* altruism, is *truly disinterested* altruism possible at all? And what part might God have played in developing these processes, especially if they involve deception? And what are the implications for the great moral dilemmas that the environmental crisis raises for Northern countries?

The problem is especially acute for practising Christians. We like to think of our religious emotions as noble and above self-interest, but 'consciousness cloaks the cold and self-serving logic of the genes in a variety of innocent guises' (Wright 1994: 275). Could that be true?

It certainly is true that almsgiving serves our feelings of self-respect, to the extent that Jesus had to warn his followers to do it quietly (Mat 6:2); the idea of loving our enemies is surely indistinguishable from revenge if its motive is to 'heap coals upon their heads' (Rom 12:20); if it gives us pain or guilt to see people in distress, helping them may be as much in our own interests as in theirs (Mat 18:31): if we serve God only in the hope of attaining heaven, we can hardly claim a disinterested motive (Titus 3:7); and so on. EO Wilson's famous comment about Mother Teresa (pointing out that her remarkable altruism was backed by her certain conviction of a heavenly reward) provoked, predictably, an outraged reaction (Grant 1993: 103), but to be fair, he did have a point.

It certainly is true that many of our moral imperatives can be understood in terms of good strategies in the game of Tit-for-Tat (p.90), and are consistent both with street justice and with scripture. For example, retribution ('an eye for an eye') helped to solve the problem, that faces any moral system, of how to deal with cheats. Sympathy would be counter-productive if it threatened group survival, so on this

149

view, long-term cultural selection has predisposed us to say, if a cheat suffers, so be it, he deserves to. But 'an eye for an eye' was a social advance in its time, because it forbade the offended party to take 'exemplary vengeance', ie worth more than the damage done. Over evolutionary time, cultural selection favouring this restrained form of retaliation conferred an advantage on the individuals and groups who adopted it, and thereby preserved the workings of reciprocal altruism without excessive social costs. Further, the extension of this score-keeping process to retributive justice in the next world may often be the only consolation for the continued injustice suffered by the poor in the present one.

When Jesus taught that we should love our enemies, he was suggesting that we should go several steps further than cultural selection, into the realm of no selection (Fig 3). He denied that the Biblical eye-for-eye retribution was a command of God, thereby removing its old authority. Instead, his teaching pointed towards a set of

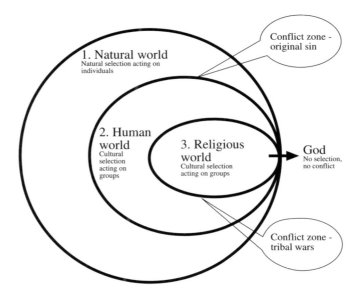

Figure 3. The interactions between the natural, human and religious worlds, and the definitions of selection and conflict.

higher ethical truths that long anticipated the conclusions of Axelrod's computer analyses of repeated games of Prisoner's Dilemma (p.89) They include: forgiveness, especially in marriage, is a good investment; it pays to help others but also to stand up for oneself; and loving thy neighbour is simply good diplomacy. Forgiveness of insults and free acceptance of cheats is not only a way of obeying Dawkins' injunction and demonstrating that we are not slaves to our selfish genes after all, but also encourages greater compassion and concern for our fellow human beings, as Jesus taught (Wright 1994: 338).

So in my view there is no question that evolutionary processes must be accepted as part of any contemporary theory of morality, including the counter-intuitive insight into the role of self-deception, but the apparent transition from evolutionary egoism to true human ethics still requires explanation.

My suggestion is that the development of true ethics is another example of the well-known mechanism by which selection can act on a character evolved for one purpose and adapt it to serve another. Whales' flippers and bats' wings were formed from reptilian feet, and mammalian ear bones from fish jaw bones, simply because evolution is a cumulative process, and the material available for new forms is determined by what has survived from previous forms. Animals are necessarily compromises of design (Eldredge 1995: 46; Williams 1996), and their ability to take advantage of the opportunities opened up for them by environmental change is constrained by history and by existing genetic variability.

The process works as well on behavioural traits as on feet and wings. One of the most convincing explanations for the evolution of intelligence is that it allows greater memory for, and skill in computation of, the complex series of remembered transactions with known individuals that underlie the continually shifting alliances and reciprocal favours of life in a primate group (Whiten and Byrne 1997). Brainier chimps are simply better players of repeated Prisoner's Dilemma (p.77). Greater skill in this is certainly rewarded by natural selection, because the alpha male of a band of chimps is the one most likely to father the highest-ranking offspring born in the group, the ones which have the best survival prospects. Once evolved and further refined, as in modern humans, intelligence was available to be applied to cultural skills, such as abstract mathematics, astronomy—and, eventually, theological speculation.

Similarly, the emotions that evolved to assist groups to maintain their cohesion by reciprocal altruism were available to be extended to what de Waal (1996: 205) calls genuine community concern among chimpanzees. It is not, de Waal is careful to point out, that these animals worry about the community as an abstract entity, more that they strive to maintain the kind of peaceful, co-operative community that is in each of their own best interests; but in evolutionary terms it is short step from there to environmental ethics.

I therefore suggest that, once evolved for related but different purposes, community concern allied with reflective intelligence, each favoured by natural selection because they enhanced the individual breeding success of our far distant ancestors just as orthodox evolutionary theory requires, became available to be refined into genuine, selfless altruism characteristic of the genuine spiritual world (Fig 3). Just as, with time and sociality (ie, repeated encounters with the same individuals), the ruthless computations of the Prisoner's Dilemma game pass from 'Always defect' to 'Co-operation pays' (p.91), so has egoism in the primate stock passed from 'Look after Number One' through 'Scratch my back and I'll scratch yours' to 'Love the Lord your God with all your heart, soul, mind and strength'. Human ethics has thereby grown beyond dependence on its animal precursors. An appropriate biological analogy has already been supplied by Keith Ward, a theologian well-enough informed on the sciences to engage in public duels with Dawkins:

> the soul need not always depend on the brain, any more than a man need always depend on the womb that supported his life before birth (Ward 1992: 152).

The reason that the hard-line reductionists do not see this extension process working is because they refuse to consider *motives*. By locating morality outside nature, Williams and Dawkins have absolved themselves from trying to fit it into their evolutionary perspective, says de Waal (1996:16).

The reductionists' insistence on leaving bits out, such as questions of purpose and motive, is a metaphysical interpretation, not a necessity imposed by the data. Their case is insufficient: it is not possible to ignore the true motives and genuine ethical struggles that do really exist, and need explanation. The solution to the puzzle, says Ward, is clear:

> if science rules out purpose by definition, and then finds that there are many things it cannot explain, one very obvious thing to think is that it is precisely the bits science has left out that we need to bring back in again . . . All we need to resist is the idea that any natural science gives the whole, complete and exhaustive truth about the real world (Ward 1992: 58-9).

The integration of multi-level selection theory and theology could in future offer suggestions on how to revise our view of Christian attitudes to creation and to the evolution of moral sensitivity in order to address questions of purpose *within* the constraints of mainstream science. For example, it is both impossible and un-necessary to deny the huge influence of genetic inheritance on human characteristics and behaviour—indeed, Jones (1996) maintains that, since genetics is often asked to test beliefs about what it means to be human, it is closer to moral and religious doctrine than any other science. But neither does this view deny the equally great influence of divine grace, however interpreted, in perfecting our nature, drawing it on into another stage unattainable by selection processes. 'Morality', said CS Lewis, meaning the true, pure-in-heart sort, worlds above the mere score-keeping that can already be observed among chimps,

> is indispensable: but the Divine Life, which gives itself to us . . . intends for us something in which morality will be swallowed up. We are to be remade. All the

> rabbit in us is to disappear . . . morality is a mountain
> which we cannot climb by our own efforts . . . [but] It
> is *from* there that the real ascent begins. The ropes and
> axes are 'done away', and the rest is a matter of flying
> (Lewis 1971).

We need the help of God's grace to be able to 'fly'—that is, to overcome the xenophobia and distrust of those outside our own group that is as much part of our makeup as is affection and co-operation with those inside it. This sort of universalism is exactly what is needed to weld humanity into a common unit capable of making collective decisions on how to deal with the environmental crisis, and perhaps the Anglican Fifth Mission Statement, and similar documents undergirded by a revitalised Christian theology of creation, might be a powerful source of it.

6. Environmentalism and the Institutional Churches

Religious organisations are not generally known to take a leading role in the environmental debate, in New Zealand or elsewhere: indeed at the level of the ordinary person-in-the-street religion is often seen as, at best, having nothing to contribute, and at worst, as being downright hostile to the natural world. In fact, there is a surprising amount of church-sponsored work on environmental matters going on behind the scenes.

I am convinced that the common opinion is wrong, that religious communities do have a vital role to play, one which does not depend on defending the outdated theologies that many people now reject. But to earn a hearing at the debating table, the churches will first have to make themselves relevant to outsiders—most especially by consolidating the contemporary insights and knowledge of many different people and disciplines into a new and coherent theology capable of undergirding their exhortations. Knowledge does not, in itself, lead to right action, but it makes the knowledgeable who do nothing responsible for disaster, as were the prophets. Therefore, the environmental crisis is both an enormous opportunity and also a strong challenge to today's churches (Houghton 1997: 110).

The first thing to consider is to what extent any modern church can fulfill the contemporary equivalent of that ancient prophetic duty. That does not mean returning to ancient attitudes—on the contrary, it means looking at them more critically in the light of modern needs. For example, we must take seriously the comment by Lynn White, that, to the extent that the fundamental values of Christianity have long influenced socio-economic patterns of western society, we cannot get out of the present environmental crisis *until* we 'find a new religion, or rethink the old one' (White 1967: 1206). Most Christian writers ignore the challenge of such a drastic re-evaluation of our faith, whether or not they agree with White's identification of Christianity as the root of the problem. But ignoring it means to continue with what Rasmussen calls

155

the 'serious mistake' of Christian theologians of the second and third centuries onwards, who thought that they could find their way to God through the contemplative mind, abandoning material reality in a preference for pure spirit, and progressively falling out of love with earth in the course of nurturing soul, mind and reason. Now,

> Experiencing the gracious God means . . . falling in love with earth and sticking around, staying home, imaging God in the way we can as the kind of creatures we are . . . If Christianity does not demonstrate a power that addresses earth's distress and makes for sustainability, its claims to be redemptive ring hollow (Rasmussen 1996: 280, 272).

Surely we must be able to find ways to do this.

6.1 The role of religion in the environmental debate

It is supremely ironic that, while so many local church communities are widely seen to be, and many see themselves to be, irrelevant to and disinterested in environmental problems, it is becoming increasingly clear that religion has an important role to play in the debate. To quote only a few writers who have already made this point:

> There is something like a religion embedded in our commitment to growth and modern industrial progress. For if it is true that we are committed to these values to the extent that we cannot live without them . . . they have become ultimate and absolute . . . in a peculiarly modern sense they have become our gods . . . But our profound belief in those objects has made us insensitive to the damage they are doing to that alienated world of nature on which in fact our lives primarily depend . . . This makes our crisis a spiritual one (Ambler 1990: 53-4).

156

The changes that are now needed in society are at a level that stirs religious passions. The debate will be a religious one whether that is made explicit or not. The whole understanding of reality and the orientation to it are at stake . . . to ignore that, to treat the issues as if they could be settled by abstract reason, is misleading . . . Getting there, if it happens at all, will be a religious event, just as getting to where we are now was a religious event. Idolatry . . . has brought us to the present crisis. Overcoming [it] is a religious task (Daly and Cobb 1990: 374-5).

Daly and Cobb refer, of course, to religion in general, not to any particular set of doctrines. Rasmussen (1996: 182) more specifically refers to Christianity as:

a particular religious worldview, and a way of life in keeping with it, [which] when armed with the powers of modern technology, will do us in.

Such comments might be thought to give aid and comfort to the church's missionary effort. Alas, the reality is quite otherwise: for most clergy, concern about the environmental crisis is practically irrelevant to their work in the missionary field.

Some authors, such as the philosopher Loyal Rue (1989: 5) admit that the church seems too hidebound and too complex to respond to such an immense challenge. Yet Rue remains convinced that 'the church is really the best place to start the work of revitalising western culture' (*ibid*, 184). He recognises the objections, but

I see nothing to be gained by giving up on the church; it will not go away. And if the church itself is not brought to make a radical departure from the past, then it will remain a formidable obstacle to any revitalisation attempts centered outside the church. The task . . . is to be accomplished not by leaving the church, but by changing it radically (*ibid*, 5).

Oelschlaeger (1994) describes his book *Caring for Creation* as a confession, an account of a conversion experience. Influenced by Lynn White, he was once strongly prejudiced against religion, but he eventually came to the opposite conclusion. He recognises that all politicians and secular experts are part of the economic system that treats the public good as a by-product, if that. Organised religion is part of the same system too, to some extent, but in its best moments it is more concerned with the welfare of the group and the collective good than is any other institution. So, he concludes, the church is our last, best chance. He states firmly that 'There are no solutions for the systemic causes of ecocrisis, at least in democratic societies, apart from religious narrative' (5).

Oelschlager builds a strong case for his view that

> The metaphor of caring for creation is literally an instrument for social transformation: it is an instrument of moral and intellectual growth . . . not a theological rule but an imaginative paradigm that might prove useful for a culture undergoing ecocrisis (*ibid*, 222).

Oelschlager acknowledges that religious metaphors are ordinarily associated with skyhooks—privileged metaphysical claims (p.74)—rather than environmental ethics, but he does not explain how to use a metaphor to persuade people to co-operate to achieve public benefits. He mentions the Tragedy of the Commons briefly but does not explore its consequences or how to escape it. He does not discuss the forces of co-evolution at all, except in the broad statement that 'the controlling memes . . . of industrial culture must be reshaped' (230). His analysis is unusual in the Christian literature for getting close to recognising the key dimensions of the problem, but his suggestions for dealing with it are impractical.

Nevertheless, as a Christian, I agree that Oelschlaeger is on the right track, a parallel (American) one to the Anglican initiative represented by the Fifth Mission Statement. On the other hand, as a scientist I am also concerned that churches should be careful how they take up their role in the debate, considering that they are latecomers to this particular field. The development of research concerning the environmental crisis is much more advanced in the secular world than in theology. During the long period when, as Moltmann put it, theological and secular thinkers achieved a peaceful co-existence based on 'mutual irrelevance' (Moltmann 1985), all the foundations of the present secular disciplines of conservation biology and environmental ethics were laid without any input from theology (see, for example, Caughley and Gunn 1996; Passmore 1980). The existing system of national parks and reserves that protects many of the most significant landscapes and natural habitats around the world—which could not be established *de novo* in contemporary conditions—owes nothing to theology. If 'the ecological crisis is . . . a *kairos* moment' for the church, as McDonagh (1994: 145) maintains—a moment of decisive challenge during which matters of great good or evil will be decided—then it is as well that others woke up to it sooner than the churches have done.

By contrast, most other current Christian literature on the environmental crisis starts from the argument that Christianity was green all along, so all we need to do to put things right is to revive the old idea of stewardship. These ideas are useful and to some extent true, but they are deeply unsatisfactory. Behind them there is a more important point that all such conservative arguments seem to miss. The various contributors to one book, entitled *The Earth Beneath: a critical guide to green theology* (Ball and others 1992) put their fingers on it. They take the view that it is not enough merely to discuss how humans have been treating the natural world: that is of course important, but behind that problem there is a deeper question, about the identity and place in nature of humanity itself.

Religious authors too often to attempt to 'domesticate' the environmental problem, by redefining it in terms acceptable to tradition. Reinterpretations of Genesis avoid the embarrassment of acknowledging the complicity of established religions in the destruction of the earth's resources, by pretending that scripture and tradition have always been environmentally friendly but have simply been misunderstood by all previous generations until now. This tactic disguises the radical nature of the challenge of environmentalist groups, and so limits its impact.

> It is far easier and safer to try to contain the challenge
> of 'green' theology within existing boundaries than to
> be open to the possibility that what it really requires is
> a complete re-think of traditional Christian attitudes. It
> is our argument . . . that most of the material that has
> been published so far has gone for the safe option, that
> of reinterpreting our existing language. We will
> suggest that what is needed is something new and as
> yet undeveloped. If Christians are to share in that
> process of development they will need to . . . be
> prepared to let go of ideas from the past that are no
> longer adequate (Ball and others 1992: 4).

This theme can be developed in various directions. Robin Grove-White argues that there are two different ways of thinking about the environmental crisis. The orthodox one stresses the role of science in understanding both the problem and the human reaction to it; the alternative account stresses the central role of human relationships and cultural contingency. The two are not, he believes, equally valid. The orthodox view, which is based on a common but seriously inadequate conception of a human being as a 'rationalist-individualist calculator' (Ball and others 1992: 24)—what Daly and Cobb (1990: 85) call '*Homo economicus*'—is itself part of the problem, whereas the environmental movement is better understood as a 'vehicle for the reassertion of public, collective values in societies whose individualism has overrun its course' (Ball and others 1992: 31).

Margaret Goodall and John Reader (in Ball and others 1992: 36) agree with this argument and expand it further: 'The real question is, "What is it to be a human being?"' They suggest that part of the pressure to consume, which is so damaging to the planet, can be seen as an escape from a deep ontological insecurity. One very common response is to retreat into the familiar past, regardless of its problems; but that poses a particular problem for Christianity, because its heavy investment in the past is no longer considered an asset by some, especially the more outspoken of the feminist groups. At the same time, Christians must combat the false optimism of New Age philosophy—that loose collection of ideas including the belief that humanity can save itself, and that it is not necessary to take evil seriously. Failing to recognise good and evil as separate, or to provide some means of dealing with personal sin, prevents personal growth (*ibid*, 57).

There is another problem, too: there is a perceived contradiction between the feeling that we should take responsibility for, and do something about, environmental degradation, and the opposite and much older belief that God is in charge of all history. In the absence of any firm teaching or leadership from within the church, most people become reduced by confusion to moral passivity. As Hall (1986: 50) puts it:

> How can we speak about God's sovereignty without undermining human accountability? . . . How shall we think of human responsibility without betraying a covert atheism? . . . Time [once] permitted us the luxury of theological debate . . . But today the whole church is confronted by the need, unprecedented in history, to . . . bring to bear on a threatened globe not only its own undivided service but also a wisdom that is sorely needed within society at large.

As the secular green lobby has been saying for years, it's later than we think.

6.1.1 Green grace and red grace

The history of Christian traditions about nature are comprehensively surveyed by Paul Santmire (1985). He shows that the exploitative theme criticised by White has from the beginning been countered by another, equally ancient but less well-known theme affirming the intrinsic value of the natural world and the presence of God in it. The implication is, says Ray Galvin, that the churches *are* guilty of aiding and abetting the environmental crisis, as White claims—not because Christianity caused the problem, but because it has neglected those elements of its own traditions that 'could have helped save the day' (Galvin and Kearns 1989).

161

Within the various strands of modern Christianity, there is such a broad range of different understandings of creation that it is now unrealistic to hope to formulate a single Christian theology of creation for the future. The best we are likely to achieve is a series of locally relevant but linked variations on the basic theme. Christians must also formulate their understanding of the natural world in terms comprehensible to people outside the churches. While the need for working together on conservation issues is so urgent, it is necessary to concentrate on what *unites* Christians with each other and with and non-Christians, rather than what divides us. Perhaps we could agree to use Jay McDaniel's useful term 'green grace' to describe the truly numinous in creation (McDaniel 1995). This concept is understood by anyone who has experienced it, religious or not, so it might be useful to recognise its value in stimulating and supporting the changes in northern life-style that global conservation demands.

However, if the churches want to make a serious contribution to the environmental debate, they must also continue with their ancient duty to mediate what McDaniel calls 'red grace' to the world, especially through the Eucharist, and do it with more love and acceptance of other points of view, more tolerance and less concern with personal righteousness, more emphasis on peace and joy and less on sin, more on intellectual rigour and less on dogma, than they have in the past.

The churches have had centuries of experience in the business of building communities, places where people find the sense of personal identity and community solidarity that drive concern about, and offer solutions to environmental issues. The changes required do not mean abandoning all tradition, but they do mean taking both creation theology and ecological science very seriously indeed. In Philip Hefner's apt analogy, the story of cosmic evolution, from the Big Bang through to the very recent appearance of human culture, provides the loom on which we can weave a new interpretation of the perennial mystery of who we are and what we know. Faith can use the loom to weave together the stories that yoke science and religion together in a new cosmic myth for our time. The loom is not the weaving, but the weaving is not possible without the loom. We can see the loom as part of the vision, constructed on purpose for it, but the weaving is done by a leap of faith, and the cloth is a celebration of nature.

6.1.2 *Game theory and environmental agreements*

A common response to any convincing explanation of the environmental crisis is: 'Now what do we do?' Religious authorities are perhaps better placed than most to point out that there are two levels of answer: one, we have to change the economic and political environments in which we live, and two, we have to learn to change our own individual lifestyles. Many mainstream churches are already active in the multi-disciplinary and much-needed critique of the current economic and political paradigms, and lists of 'What you can do as an individual' are widely available from green organisations and academics (Wyman and others 1991), but nearly all such efforts fail to take into account the problem that knowledge does not lead to virtue in a society in which ignoring virtue is a profitable strategy. Exhortation, however authoritative and well-intended, cannot make a difference if it ignores what game theory can say about how people make personal decisions, especially those involving resources.

Throughout their formative years, our ancestors lived in small, homogenous groups in which every individual was familiar to every other one, and all had a stake in the survival of the group. Where all members of a group have interests in common, co-operation is beneficial to all, and constantly reinforced by evolved social attitudes forged and maintained by deep-rooted habits of reciprocal altruism available only to other members of the same group. Competing groups were independent of each other, and generally hostile (Campbell 1975). But modern human society has become very diverse and complex, so the ancient rules generate tensions both within and between the main groups (eg, nation-states and political or trading blocs).

National and international agreements on ethical issues depend on co-operation between diverse groups or subgroups with different interests, hammered out against the grain of the built-in human tendency to try to score points against members of other groups (Alexander 1987). The key problem in environmental ethics is to persuade individuals voluntarily to extend meaningful co-operation, and enough of it, to others *outside* their own familiar groups, even though that defies the ancient 'default setting' (p.111) of human nature

(Heinen and Low 1992). Game theory explains why that is so difficult: caution, lack of trust, suspicion of concealed motives, unwillingness to allow the other side any advantage, intolerance of foreign free-riders combined with tolerance of free-riding at others' expense, have been the best way to negotiate with other groups for virtually all human history until now.

The game theory approach applies as well within nations as between them. For example, consider the matter of litter in and around any modern city. The citizens as a group would benefit if *everyone* were responsible about litter; if no-one ever dropped rubbish in the street or discarded condoms in the bushes, or threw burning cigarette butts out of their car windows, or dumped trailer-loads of junk at the side of the road. The city council would save the thousands a year it now has to spend on cleaning up assorted messes in public areas, the parks and central city business area would be more pleasant to look at and walk in, the risk of disease and damage from pests would be minimised, and everyone would benefit for the expenditure of minimal personal time and self-discipline. But it does not happen. Why not?

The decisions made by any one member of a group depend substantially on the decisions made by the other members. Table 3 sets out the options in the form of a game in which I play as an individual against the rest of the citizens as a group. Since I cannot know all of them individually and am unlikely ever to meet them all as a group, this is the equivalent of a one-off Prisoner's Dilemma game (p.87). For the moment, it assumes I have no strong principles either way about litter, but am merely objectively calculating my options. The table shows that, if everyone obeys the rules the city is clean, though everyone shares the cost of personal self-discipline. My personal result is better if I defect when everyone else is co-operating, because I then enjoy a clean city without having to change my own habits. But if I co-operate and others don't, I risk having disciplined myself for nothing, in which case I might as well save myself the trouble and contribute to the mess along with everyone else. So whatever everyone else does, my best strategy is to carry on as normal. The result will be a local version of the *Tragedy of the Unmanaged Commons* (p.47).

In real life the people who live in any given city are not a single,

Table 3 Game theory analysis applied to the question of why not everyone living in a city will co-operate to achieve a public good.

The aim is to reduce the cost, health hazard and aggravation caused by litter in the centre of a modern city. The decision payoffs are calculated with reference to any one individual ("what I do") playing against the rest of the population as a group ("what you do"), ie, the table is to be read from the left in rows, rather from above in columns. Co-operation is defined as always using the council litter bins or taking my own litter home; defection as the opposite behaviour. The same argument can be extended to other forms of "litter" such as graffiti and vandalism. The scores allocated to each outcome (in brackets) are arbitrary but in about the right order of desirability.

		What you do	
		Cooperate	*Defect*
What I do	*Cooperate*	I enjoy a very clean city at the cost of some self discipline (3)	I suffer the visual pollution even though I have been self-disciplined (0)
	Defect	I enjoy a mostly clean city and save myself the trouble of self discipline (5)	I suffer the visual pollution, with the small consolation of not having to be self-disciplined (1)

homogenous group. There could well be a large difference in attitude between those who inhabit, for example, the up-market suburbs versus those who live in the poorer areas, or between those who own their own property and pay rates versus those who pay only rent. For the sake of this argument, we could call the two groups the Blues and the Reds. Many of the Blues would perceive themselves as stake-holding members of the community with an interest (at least in theory) in the welfare of the city and its community. They are sensitive to visual pollution so would benefit from greater community co-operation over litter in public places, and most would never dream of dropping litter in the street. By contrast, the Reds are more likely to see themselves as having no stake in the community and nothing to gain from putting themselves out. They are often insensitive to mess, unthinking or unmotivated about civic responsibility, and have nothing to gain from a reduction in litter since they are not offended by it.

Controlling the litter problem by citizen's responsibility alone would require full co-operation between all sectors of society. But co-operation between groups depends on the degree to which they share common interests, whereas it is more likely that the Blues and Reds will have different attitudes to litter, which will affect their weighting of the options. Blues are much likely more to value co-operation on principle, even at personal inconvenience; Reds are much more likely to defect - and not only by failing to co-operate. Reds who feel they have no stake in the community because their interests and concerns are ignored by the other stake-holders may also use litter, graffiti and vandalism to damage public amenities that Blues value. The local variation in proportions of Blues and Reds will produce different combinations of outcomes for the litter problem in different suburbs.

The problem of litter in public places is a simple one but similar in principle to many others faced by environmental activists everywhere. The cause is often seen to be ignorance —a failure of schools and parents to teach civic responsibility to children. In that case, the appropriate action would be an effective public education campaign, but this analysis casts doubt on that strategy. The Blues, who are already responsible about litter, put a high a value on co-operative behaviour in all circumstances, and the Reds put so low a value on it that they are not likely to incur the cost of self-discipline merely in response to exhortations. Clearly, defection is the safest option for the least privileged parties in a structured society in which some members have a lot less to gain from co-operation than others. No-one will support a society, or help keep its streets clean or listen to its elected representatives or help save electricity during a national shortage, unless they feel part of it, which is why general appeals for co-operation on matters such as civic responsibility and public consideration *never* get a one hundred per cent response.

On the global scale, the same logic applies. The present urgent need for groups to trust each other and work together at unprecedented levels does not alter the way we think, itself forged over thousands of years' experience of a very different world. The tensions and bickering at the Earth Summit in Rio, and the high-profile defections by US and Australia from the Kyoto Protocol, are proof of that. The differences in wealth, ethnicity and social status between the participants in such international negotiations are huge, and the processes by which social inequity undermines co-operation are all too clear.

On the other hand, the churches are not much better than secular organisations at promoting naturally-evolved co-operation *within* groups, which is part of human social life everywhere. There is a sense in which every congregation is a social unit no different from a bridge club, and personal tensions are rife in both. But churches could be in a better position than most secular organisations to help overcome the barriers to co-operation *between* groups. Those barriers include racism, moralism, local and personal self-interest, lack of confidence in others' good will, and many other evils that Christianity officially abhors but actually often practices itself. Given the right circumstances, the churches do have the capacity to mobilise broadly humane sentiments across society in order to make a coordinated contribution to the public debate—for example, during the Hikoi of Hope demonstrations throughout New Zealand during September/October 1998 (p.173). Elsewhere, the churches have made important critiques of public affairs in the Phillipines, Brazil and US, always working towards peaceful revolution and playing down old antagonisms (Martin 1997: 220). A church that truly lives up to its founders' values, that deliberately rejects the old moralistic, doom-laden attitudes and concentrates on its real work of transforming people by grace, *could* make a difference.

6. 2 The debate about stewardship

Christian theology has long squirmed under the repeated criticism, mostly but not only from outside the church, that Christianity is irredeemably instrumentalist in its attitude to nature, and therefore unqualified to address the issue of caring for creation in terms that mean anything to modern ears. On the other hand, some theologians and philosophers such as Passmore (1980), Attfield (1983), Hall (1986), RJ Berry (1991) and Wendell Berry (1993) have sought to show that the idea of stewardship is a neglected but legitimate strand in Christian tradition. According to these and other authors, the arguments required to meet the modern need can be found within existing traditions.

By contrast, others such as Primavesi (1991: 107) retort that the concept of stewardship is still exploitative and unecological, since stewards seek to optimise profits for themselves or their bosses.

Primavesi criticises the idea that 'Christian hope is the most dynamic stimulus to work . . . for the stewardship of all creation *for the benefit of all men and women*' [her italics]. She objects to the inappropriate 'hierarchical thinking [that] encourages man to believe that he is "above" or "in charge of" his ecosystem. In truth, he now seems to be in charge only of its destruction' (*ibid*, 94). Many other writers have made the same point, eg McDonagh (1994: 131) and Hallman (1994: 6).

But resistance to any attempt to change the deep-rooted anthropocentrism central to Christian theology is strong. Rasmussen (1996: 230-36) reports that, at the 1991 Canberra Assembly of the WCC, three paragraphs of a draft document referring to humans as "one species among others" were deleted because too many delegates objected that it offended the Biblical dignity of humans and the divine calling to stewardship.

In the concept of intrinsic value we find a welcome congruence of ideas between theology and secular philosophy. 'Deep-green' philosophers have for some time disputed the traditional Christian instrumentalist view of nature, arguing that nature should be valued for its own sake (Sylvan, 1992). Now theologians are rediscovering the same idea in the Genesis account of the covenant with Noah, which included the animals and all life (Birch, Eakin, and McDaniel 1990: 277). Nevertheless, stewardship remains the dominant paradigm of environmental management, both in Christian (RJ Berry 1991; Peacocke and Hodgson 1996) and in secular (Attfield 1983; Passmore 1980) writings. The Anglican statement to the UNCED conference in Rio (Berry 1993a: 263-4) emphasises both the human responsibility for stewardship of nature and the contemporary convergence in religious and secular ideas of it. The John Ray Initiative, recently established as a limited company and registered charity by Christians in Science, states that its purpose is to promote environmental sustainability and responsible stewardship.

The main problem with all this is that good stewardship requires the weighting of all interests, which is especially difficult in any system in which the managers are also part of what is being managed. Weighting of interests is possible to calculate in economic terms, as in the interdisciplinary study reported by Attfield (1996), and as advocated for

New Zealand by Hartley (1997). Unfortunately, this approach has the serious consequence that importance values, which cannot be quantified, must therefore be underestimated, and so 'much of what those who care deeply about the environment want to say will not be communicated clearly' (Grove-White and O'Donovan 1996). Future progress in designing conservation policies may depend on the extent to which we can forge closer co-operation between Christian, secular and business viewpoints.

The final document from the WCC World Congress at Seoul sticks to the traditional Biblical perspective, that creation is separate from God. It declares that 'because creation is of God and *the goodness of God permeates all creation, we hold all life to be sacred* '; nevertheless humans are 'created in the image of God with a special responsibility for the rest of creation'. The WCC thereby acknowledges an intrinsic value for creation, which is a considerable step forward. But, as one would expect from a Bible-centred organisation, the WCC at the same time distinguishes itself from deep ecology, which denies any special role for humans. It could be argued that the case for protecting the environment can be best argued by an international authority which recognises both secular and transcendent values, such as a revitalised Christian Church.

The trouble is that the Biblical bases of key concepts such as the place of humans in nature or the ideal of stewardship are very much more complex than is usually recognised. All three possible human attitudes to the natural world (mastery over, subjugation to or harmony with nature) are present in the Old Testament. They are integrated into a single world view shared by all Israelites, though most individuals would have preferred one orientation over the others, depending on their sociological backgrounds (Simkins 1994: 171). Palmer (in Ball and others 1992) demonstrates the difficulty of defining a clear-cut Biblical pedigree for any general principle of stewardship, or of transferring one of the several scriptural possibilities without distortion into the modern world. More importantly, she questions whether any possible reinterpretation of the term is adequate or appropriate for the very different view of the relationship between humans and the natural world which is required in a modern, as opposed to a feudal, society.

For example, in both the Old and New Testaments the concept of stewardship appears in the context of a servant put in charge of the property of an *absentee* master. The application of that idea to humanity caring for creation carries the implication that God is far away, no longer present in the world—and further, that creation is an inert substance of little interest to God except as property to be managed in order to gain a good return as an investment. Neither idea squares with the equally ancient scriptural notions of God as delighting in creation in and of itself, long before the arrival of humans, or of the immanence of God in every form of life so intense that the whole of creation can be called, metaphorically, the body of God (McFague 1993). Worse still, the stewardship metaphor is responsible for the old idea that nature is somehow incomplete until it is put to use by humans. Hence the first settlers in US and Australia justified their taking land from the indigenous people who were not engaged in agriculture or visibly using the land in ways that they supposed God had intended it to be used (McDonagh 1994: 131-2). Such incidents suggest that there is some truth in the philosopher Stephen Clark's description of the stewardship ethic as 'licenced banditry' (Clark unpublished.[1])

An alternative approach to stewardship is advocated by Hall (1986: 89ff). First, he draws a distinction between two primary interpretations of *imago dei*, the idea that humanity was created in the image of God.

1.The first and best-known, *substantialist* interpretation defines the image of God as those capacities or endowments of human nature that are distinct from other natures—especially rationality and freedom of will, both seen as the 'stamp of the maker'. This exaltation of rationality was foreign to Hebrew thought, but nevertheless it had a strong influence on later interpretations of Genesis in the Hellenistic world. It led to the claim that, since fallen humans were still rational, the effect of the Fall was to distort but not to destroy the image of God in humanity.

2.The second, *relationalist* interpretation holds that *imago dei* is not a quality of nature located in us, but a relationship: 'to be *imago Dei* does not mean to have something but to be and do something: to image

---

1.   Verbal comment contained in a paper delivered to a joint meeting of the Science and Religion Forum and the British Ecological Society, at Hoddesdon, UK, September 1996.

[verb] God', to be turned towards, to be in relationship with God. The effect of the Fall was that the image was totally lost, and the inevitable result was death. We can reflect God when we are turned towards God, even though we are small in comparison (a dewdrop can still reflect the sun); but when we are turned away we cannot.

> In our state of estrangement . . . we no longer image God, not because we have lost some inherent quality of our creaturehood but because we are literally *disorientated* (Hall 1986: 106).

There are profound implications for religious environmental activism in recovering the relational understanding of *imago dei*. It argues against the exaltation of rationality and responsibility, which are among the besetting sins of secular environmental economists such as Hartley (1997). By implication, it therefore also argues against the consequent denigration of other, irrational and irresponsible creatures, which in the worst cases have been taken to include indigenous people. More important, it allows a critique of the ambiguous benefits of human reason. That is needed, says Hall, in words that go to the heart of ecotheology, because:

> With our rationality in full cry, we have now created a technological society in which it is almost impossible to live like truly human beings . . . we have built a *civitas terrena* poised on the edge of oblivion (Hall 1986: 111).

Hall sees a deep paradox in the idea of a Christian stewardship that makes no connection between the confession of faith that Christ is both Lord and image of the invisible God (Col 1:15) and the example he set of what lordship implies, i.e. humble and loving service. If we are to be conformed to his image and be incorporated in his body, we have to recover the concept of servant-hood too, and 'such a belief ought to transform the whole idea of human domination [=lordship] within the realm of nature'. This sort of dominion can only mean stewardship, ultimately interpreted as self-giving love. That in turn would transform our values, from *dominion over* to *solidarity with*, from *power over* to *service to*.

171

How could this long-established alternative interpretation of stewardship have been so long ignored, and still be so unpalatable?

> The answer is, of course, that the lordship of Christ was itself soon transmuted by imperially placed Christianity into something very different from the actual testimony to the life of the Lord given by the Biblical writers. Jesus was invested with all the trappings of earthly monarchy by an adoring church . . . the radical model of authority and majesty that Jesus as Lord actually embodied, with its intrinsic but unmistakable polemic against power, was all but lost to evolving Christendom . . . The challenge issued to the Christian movement today by the crisis of the biosphere, namely, that it develop a more adequate theology of nature, is thus at the same time a challenge to develop a more authentic Christology . . . we have to recover a Christos whose lordship is [more human and] vastly different from the magisterial model preferred by empirical Christianity (Hall 1986: 185-7).

Hall's analysis thereby picks up the damning critique of Constantinian Christianity made by Kee (1982), and illustrates why meeting the environmental crisis is such an enormously threatening and far-reaching task for traditional theology. It simply has too much to lose. Over the long term, it illustrates why we should not simply take the easy way out and fall back on the stewardship metaphor, inseparable as it is from the mastery-over-nature syndrome, as urged by many influential voices in the churches today. But while the existing churches are in the process of rethinking the doctrine of creation and the place of religion in society, which will take more time and will than any of them have given the job so far, the stewardship metaphor is a start.

6.3 Implications of church-led social activism for theology

The churches in the North have not yet come to grips with the degree to which Christian theology and tradition are associated in the public mind with the economic strategies and attitudes that have dominated our societies since the Industrial Revolution (Hallman, 1994). A recent and clear example of this arose during the debate stirred up by the Hikoi of Hope in October-November 1998.[2] The moral authority of the Anglican bishops of New Zealand to lead the Hikoi was to some extent undermined by accusations of hypocrisy. How could the Anglican Church, itself a large property-owner charging market rents to its tenants, presume to criticise the Government for doing the same? How can Anglicans decry poverty among the people while at the same time spending huge amounts on maintenance of their own buildings, including $20 million on a new Cathedral in Auckland?[3] The huge amount of good done by the Anglican Social Services, Anglican schools and similar organisations gets little balancing publicity. The Established Church's institutional position and history are very much mixed blessings to those of its members attempting to implement its Fourth and Fifth Mission Statements (concerning social justice and environmental protection, respectively) in the contemporary world.

Therefore, any programme involving Christian institutions in political action in modern society must start by seriously considering the churches' own position in society and in business. Christianity started small and strongly democratic; all important decisions were made by discussion and/or vote, from the replacement of Judas Iscariot to the choice of the first deacons (Acts 1:26, 6:5). But after Constantine, the church came to be organised by a hierarchical system of government which paralleled the administration of the state (Kee 1982: 167). Primavesi (1991:100) has pointed out that the consequences of this hierarchical (and, inevitably at that time, also patriarchal) thinking were

---

2. The two branches of the Hikoi (march) began at opposite ends of the country and converged on Parliament in Wellington, to publicise the concern of the Anglican Church and a host of other organisations about Government policies on social and environmental matters.
3. Former Prime Minister David Lange, interviewed by Paul Holmes on TV1 on October 1 1998.

to reduce the status of women and of the natural world, and thereby to contribute to the environmental crisis of today. The Roman Catholic Church is still organised on the model of Constantine as God's representative, sitting above the council of bishops. It is a model that has served the Church well, but it does not spring from Christian values, which are grounded in the revelation in Christ that God does *not* act as do the kings of the earth (Mat 20:28). But, by what Kee calls a 'triumph of ideology', Constantine's view prevailed, and ever since then, the established church has been associated with the ruling classes.

That strategy been very successful for centuries—at least, if a successful church is defined as one having power and influence in society—but in the very different world of today it is becoming an obstacle in the path of the church's real mission. Hall (1986: 231) puts it bluntly:

> Where churches have sufficiently de-Constantininized themselves to become distinguishable from the dominant classes of their host societies, they are able to overcome much of the harm . . . [and] join in international quests for peace without succumbing to the ideologies of politically-motivated peace movements; together with other human agencies, they can achieve humane, economic, ecological, human rights, and other goals, and help build trust between mutually suspicious and competitive governments and movements . . . [the church's] greatest challenge is to extricate itself from the Constantinian quest for power through proximity to power, and to elaborate a gospel and ethic that is truly world-affirming.

The Christian understanding of God originally emphasised Jesus' message of special concern for the poor, but history has shown that when Yahweh is detached from his historical identification with the history of oppressed peoples, he soon becomes the god of their rulers (Theissen 1984: 72). Until Christians accept that, and work to change the public perceptions that link the churches with those who have had more

174

to do with implementing current economic policies than with suffering from them, our impact on the contemporary social and environmental debate will remain limited.

6.4 Meeting the approaching catastrophe

No informed parties to the environmental debate believe that it is still reasonable to expect Northern life to continue as it is now into the indefinite future. Secular environmentalists are finding that, despite decades of rhetoric, direct action and a few battles won (such as cleaner water and more recycling in rich countries), not a single major aspect of the global ecological crisis has yet been reversed. The 1990s was identified by many (p.2) as the last critical decade for implementing the changes that might save the world. The Final Document from the Seoul Convocation on JPIC stated that 'Unless far-reaching changes are made *now* [1990], the crisis will intensify, and may turn into a real catastrophe for our children and grandchildren' (Niles 1992: 165). Now the 1990s have gone, yet still, little has changed.

The mental image that comes to my mind is of a bus marked 'Free-market economy', which is hurtling along a road through a city called 'Utopia' (Fig 4). The driver wears earmuffs, dark glasses and a broad grin, but his attention is more directed to what is going on among the passengers than on the road ahead. Blazed across his shirt is the single word 'GROWTH'. Inside the bus, the seats all face backwards, and are occupied by dozens of representatives of social and environmental agencies arguing, drinking, reading, playing cards, or sleeping—but none ever looks out of the windows of the bus, and if they did they would see only where they have been, not where they are going. (As (Wolpert 1992: 171) points out, we enter the future backwards). Along the road sides are a few mouse-sized protestors holding up placards saying DANGER AHEAD or STOP! but they are ignored. Their carcasses litter the road behind, but are too small to be noticed. In the distance the road ends abruptly at an immense cliff face labelled 'Collapse of western civilisation', and there is no ambulance at the bottom. For a little while longer, members of the present churches can choose whether to place themselves among the inattentive passengers, or among the mouse-

175

sized protestors; members of future churches seem unlikely to be able to avoid sharing the bloody mess at the bottom of the cliff.

It seems to me that the approaching crisis is about to put the churches, together and separately, in a position unique in contemporary and future society, for three reasons.

The first reason is that, while our options are still open, Christianity's deep roots in, and influence on, both South and North put it in a good position to help mediate in the international conflicts between the haves and the have-nots which have already changed most Northern societies out of all recognition, and are bound to accelerate in future decades. As those tensions increase, we might recover the full meaning of those subversive words of the Magnificat, which in our day are deadened by repetition and inaction: 'God has put down the mighty from their seat, and exalted the humble and meek; God has filled the hungry with good things, and sent the rich away empty'. If the Anglican Church is serious about its Fifth Mission Statement, and other churches likewise about their equivalent intentions, they need to recover their pre-Constantinian prophetic function, to comfort the afflicted and afflict the comfortable.

The second is that intelligent action on the environmental crisis depends on our developing some way to *interpret* the role of mankind in first causing, and now, maybe, managing the troubles we now find ourselves in. We need a credible theology of humanity capable of taking a loving but coolly realistic view of human nature. Rational Christianity can integrate the scientific account of the evolution of true human morality from its amoral animal beginnings with its own understanding of the extremes of human behaviour, from the seldom-attained heights of sanctity to the all-too-often plumbed depths of evil. This perspective can offer realistic answers to questions such as: Why do environmentalists' exhortations to responsible living so often fail? Why is earth care not the first concern of everyone? Much more than any secular environmental agency, the churches have long experience and deep understanding of every form of the multiple sins that are now arguably threatening the survival of the human species.

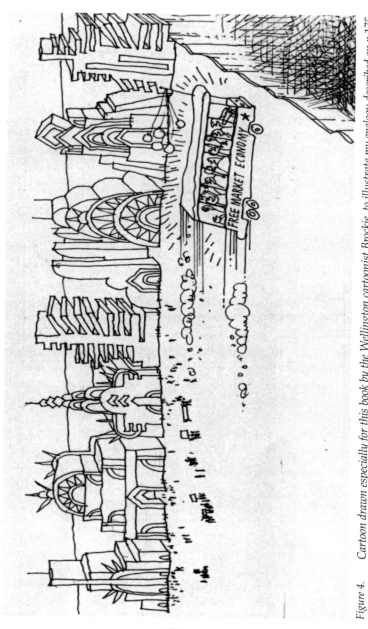

Figure 4. Cartoon drawn especially for this book by the Wellington cartoonist Brockie, to illustrate my analogy described on p.175.

> Humans, in covenant with God, need to impose a
> moral order on largely amoral nature, including their
> own . . . [they] are co-participants in an ongoing
> transformation of creation so that the future might yet
> be better, from the standpoint of justice, than the past
> (Rasmussen 1996: 244).

The third reason is that, alone of all social organisations in the world today, the catholic church represented by its various modern forms has lived through two previous experiences of complete social disintegration, once during the sack of Jerusalem in 70 CE, and again during the fall of the Roman empire and the centuries immediately following. The loss of Jerusalem was certainly the end of the world for Jews, even though the rest of civilisation was unaffected. However, the impact of the collapse of the central rule of Rome on life around the Mediterranean in the fourth and fifth centuries must have been equivalent to that which a global collapse would have on us now. At that time, the church survived and even expanded into new areas, and where the clergy were able to influence the new secular leaders of the day, they were often able to persuade them of their religious and moral responsibilities as rulers. And even where such influence was not possible, bishops or monasteries could often provide in their immediate areas a measure of charity, justice, education, and above all spiritual leadership (McManners 1992). Secular organisations based on power politics seldom long survive tumultuous social change, as game theory predicts (p.90) and as the recent history of events in eastern Europe has been amply confirming. But in the Rome of the past, the Poland of the present and, maybe, an England or New Zealand of the catastrophic future, authentic, realistic, suffering and caring church communities will thrive and continue their vital support work, as they have done before.

Whether Jesus intended to found a new religion or not—and the argument of Sheehan (1988) is that the early church distorted the message he came to give Israel about the Kingdom of God into a totally different message about Jesus himself—his church did survive the collapse of Rome, at least partly because it focussed the grace of God on, among other things, the vital role of community life in establishing the local co-operation that helps people face massive social transitions.

By the grace of God the churches still have access to sources of strength and meaning that are independent of the present world order. If our best efforts cannot actually prevent the massive threatened damage to the natural world and to society, the churches will be in a key position to help alleviate the consequences. When that future crisis comes, the churches should not only be able to stand, but they should also be able to combat despair with Christian hope by building genuine community among the remnants, and to take a longer view of events than can those whose concept of ultimate reality is more materialistic. Herman Daly, a level-headed, religious and knowledgeable man formerly working in the thick of the action at the World Bank, summed up an appropriate attitude for the informed Christian contemplating a bleak view of the distant future as follows:

> I make a distinction between optimism and pessimism on the one hand, having to do with the betting man's expectation based on evidence; and on the other hand, hope and despair, the existential/religious attitude one imposes on the situation. I think one can be hopeful and still pretty pessimistic (quoted by Athanasiou 1996: 56).

Rational faith supplies ultimate sanity in a threatening world, a defence both against despair and against apocalyptic unrealism. Daly would no doubt feel at home in the church of the future, which Hall (1986: 160) envisages as:

> no doubt numerically reduced but also less beholden to worldly powers and superpowers, [it] will prove the one inter- and trans-national movement capable of upholding and communicating a vision of world community that is not just another cloaked ideology of empire.

Daly and Hall both know that it is possible to trust in God without denying tragedy—and so does anyone who has lived through any sort of personal trauma. We need only read the book of Lamentations, written after the destruction of Jerusalem by the Babylonians in 587 BCE, to get a foretaste of what a total social collapse would be like for us. But the Jews never abandoned their cultural identity or their hope in God; they survived, and in time they returned to rebuild Jerusalem. The history of Judaism shows that to trust God is not to opt out of reality, but to choose genuine hope rather than either cheap confidence or fatalism.

If the worst comes to the worst, and the gloomy predictions summarised on pp.2–4 turn out to be true, will that necessarily destroy our concept of the loving purposes of God? Not if a more humble church of the future has a believable theology of a patient, loving and suffering God who gives us freedom to determine our own goals and then shares the consequences with us. Not if that church's own life and liturgy helps the surviving people to recognise that humanity will have brought tragedy on itself, and cannot attempt to escape responsibility by blaming God for it. And not if, having done what is possible to change what can be changed, individuals accept what cannot be changed and make the decision to live differently while they do live, to live as if life matters while they have it, and to live with integrity in the light of the brutal reality that will define that not too distant world (McFague 1993: 208). Spiritually aware people have lived with honour through the many smaller-scale catastrophes of the past, and those of the darkest future will still have the option of doing the same. The only option that will not then be available, as people survey the effects of human activities on a ruined planet, will be a church which continues to teach its new members that humanity is the crown of creation. And the final loss of that option will be no bad thing.

But I believe there is a chance that vital, compassionate and rational Christianity, working together with other major world religions, can help defeat that dismal prospect. Churches which really grasp that vision can make a significant contribution to the global effort to save from destruction our beautiful blue planet, our meeting place with God, our habitat of grace.

7. General Conclusion

The environmental crisis is a *moral* issue, because it concerns the process of reaching communal decisions about the allocation between competing groups of common resources in short supply, such as finance for conservation, access to forests, fisheries, clean water, clean air, etc. The relevant context for understanding the moral dimensions of environmental protection must include contempory biological and philosophical knowledge, because we need to understand what decisions are required, as well as the reasons why so many people ignore the interests of the environment on which we all depend. It might also include input from moral agencies such as the various forms of organised religion.

In this book I have explored some ways in which the insights of secular science can help Christians make a constructive contribution to the secular debate. From economics we can learn why the current free–market model is so subversive and why management of environmental common goods is so difficult (p.45); from game theory, why the personal restraint for which green activists plead is often not rational, except within the context of stable community life (p.71); from primatology, what are the evolutionary and social bases of morality and intelligence; from anthropology, how the combination of intelligence and socially–mediated morality as a conditional strategy has coaxed our primate and tribal human ancestors over time from rampant xenophobia through cautious trading of goods and ideas through to the philosophical analysis of true human ethics (p.92). The biological account of the origin and general operation of morality is very different from the theological and philosophical one, but is backed by a large and growing body of empirical evidence. It must be considered in any moral exhortation intended, like those made by official religious organisations such as the World Council of Churches, to be credible to non–Christians. The Christian understanding of true altruism (charity) remains a matter that goes beyond biology and into the realms of grace (p.125).

An updated Christian theology of creation could arm the Church to play a leading role in the environmental debate (p.155). Christian theologians should be among the very first to respond to EO Wilson's call for consilience between all branches of learning (Wilson 1998), since the unity of all knowledge is an ancient belief of the Church. Rational, passionate and updated Christianity could make a real contribution to developing some solution to the environmental crisis, to the extent that any solution is possible: otherwise, it will remain, as in the past, part of the problem.

8. Summary

The facts of the approaching global environmental crisis are widely agreed by secular researchers, and their general trends, at least, are not in dispute. Respected estimates give us between forty and seventy years to reduce the total environmental impact of human activities, or face widespread disintegration of the natural and social worlds. Most environmental impact, and therefore most of the responsibility, lies with the wealthy industrialised countries of the North (including New Zealand). This book takes these facts as read, and considers how some of the possible interactions between biology and theology might help the churches make a contribution toward the contemporary environmental debate.

It is widely agreed that the appropriate action is necessary *now*, and secular activists have drawn attention to many important but curable environmental problems. But exhortations to modify the environmental impact of the unsustainable Northern way of life are generally met, either by token changes in a few superficial regulations that leave the underlying trends to continue as before, or are ignored altogether, largely because few Northerners are prepared voluntarily to reduce their standard of living.

The root problem is one of values, which implies that value-based religious institutions ought to have a part in the discussion. On the contrary, the churches have so far offer scant leadership, because they are too often scientifically uninformed and distracted by internal issues. If they were doing their real job, they would be concentrating less on internal politics and more on the central problems of life, of which the global environmental crisis must rank as one of the most urgent.

Science may be understood as abstract speculation about the nature of matter, as distinct from technology defined as the practical management of life in the material world. Likewise, theology may be understood as abstract speculation about the nature of God, as distinct from religion defined as the practical management of life in the social and spiritual world. Under these definitions, this book is mainly about the *religious aspects* of the environmental crisis and about the *visionary faith* and *leadership* that the churches must find in order to face it.

There are important distinctions between faith, religion and doctrine. Faith is personal and ultimately derived from the original apostolic *kerygma*, while religion and doctrine are social, cultural, and intellectual and expressed in the mass of traditional *didache*. Religion and doctrine can be challenged and adapted in the light of secular knowledge without threatening faith, just as a hermit crab can grow a new exoskeleton and change its shell without affecting its personal identity.

Barbour's four-fold classification of the types of interaction between science (here, mainly biology) and religion provides a useful framework for interpreting the history of religious attitude to nature. I follow Barbour and Peacocke in advocating critical realism as an appropriate means of searching for ways of integrating their different insights into truth. The basic assumption of critical realism in both science and theology is that *existence* precedes *theorising*.

Over the last fifteen years the Justice, Peace and Integrity of Creation programme of the World Council of Churches has begun to examine the issues raised by the environmental crisis, and has stimulated parallel studies in other churches. The Anglican Fifth Mission Statement (adopted in 1990) states that it is part of the mission of the Anglican Church to 'to strive to safeguard the integrity of creation and sustain and renew the life of the earth'. The Statement is already raising profound questions about many ancient traditional doctrines.

The affirmations of the 1990 JPIC World Convention in Seoul, and the various declarations made at Rio in 1992, have been widely debated and have stimulated much scholarly literature. But their impact on religious activities and attitudes so far has been small, despite the urgent need for action. Religion and philosophy have difficulties in explaining this apparent blindness, but biology and evolutionary psychology can suggest reasons which need to be taken into account.

The Christian doctrine of creation is derived from the Old Testament, in which the primary emphasis was on the continual dependence of all creatures on God rather than on the origin of life. God was understood to have established a single code of laws by which nature operated, which were in principle rational and understandable by humans because they share to some extent the rationality of God. There was a range of interpretations of the natural world, but they did not include *Creatio ex nihilo,* a late addition dating to about 200 BCE.

The three main themes of the historic creationist tradition, asserting that the universe reflects the goodness, rationality and freedom of God and therefore creation itself must be good, rational and contingent, were in due course incorporated within Christian faith and are entirely compatible with science.

For a relatively short period, from the 16th century until recently, science and religion have regarded each other as irrelevant, but their ancient compatibility is being rediscovered during contemporary adjustments to creation theology. Acceptance and understanding of this compatibility is a keystone to the future success of church-led environmental activism, and the related problem of combatting literalism.

Caring for creation is a recent concern for most churches, but is not to be dismissed as a ploy to bolster sagging congregations. Environmentalism is a recent concern for all other long-lived institutions too, and each one should bring to the debate its own particular insight. For Christians that includes the trinitarian belief, that a material life thought worth sharing in person by the Word of God is thereby affirmed as worth our greatest care.

To the Hebrews, the concept of the integrity of creation referred to the fact that creation had never disobeyed God. Only the humanised parts of nature were affected by the Fall, and the hostile parts of it (thorns, thistles etc) were only conveying God's judgement on humans. The idea that Adam's sin corrupted the whole universe is a much later addition, incompatible both with all Patristic traditions and with science.

The Biblical idea of the integrity of creation is therefore a theological one, including static concepts of wholeness, purpose and 'flourishing'. It has very little relevance to modern conservation science, which understands nature as dynamic and competitive. But many proponents of both disciplines agree that there is a moral dimension to our understanding of nature, which raises ancient and deep questions about humanity's place in the universe.

Within any cultural group, traditional understanding of nature has always provided a template for moral and social behaviour, bound together in universally accepted myths. During the period of the historic estrangement of science and religion, new technical knowledge has destroyed the links between facts and values and undermined community loyalty and responsibility. In contemporary times, the breakdown of the link between the prevailing cultural story and the world as described by the sciences has decoupled reality from morality. At the same time, the parallel breakdown of community has loosened the old systems of reciprocal altruism.

The combined effect has been to weaken both religion and community spirit, and to allow free-market economics to dictate the direction of development of society according to its own supposedly neutral but actually destructive values. Hence, churches taking the environmental crisis seriously must consider relevant aspects of economics, social behaviour and game theory as well as the more obvious subjects of ecology and theology.

In the prevailing free-market economic world model, humans are assumed to be rational, independent agents making decisions in their own best interests regardless of the common good. Many environmental consequences of their private decisions are classed as 'externalities'. Commercial operators are encouraged to privatise benefits and socialise costs, each to the fullest extent allowed by society. Hence vital group-level issues such as the capacity of an ecosystem to sustain life, or the network of personal relationships within a human community, are insufficiently accounted for, and very often damaged.

When Adam Smith proposed his idea of the 'invisible hand', by which success in private business automatically benefited the community in general, he assumed that private business decisions were always restrained by shared community values. That no longer applies, if it ever did.

When resources are widely scattered, or human communities too large for personal communication or unstable, as in modern society, co-operation for the common good is jeopardised, for reasons well modelled by various forms of game theory such as the Prisoner's Dilemma: altruism is too risky, because it never pays to restrain personal interests unilaterally or to co-operate with strangers.

However, in a healthy and stable local community, the conclusions from game theory are different: repeated rounds of Prisoner's Dilemma played with the same partners can define the conditions under which co-operation is the best longterm strategy, and the result is the development of complex systems of reciprocal altruism within small groups. These conditions can be observed among the higher non-human primates, and, in conjunction with gene-culture co-evolution, were important in the extension of reciprocal altruism into systems of morality among the earliest humans.

Unfortunately, most important environmental decisions have to be taken in the context of large human groups, which naturally compete with each other. So in general, environmental ethics has to be taught in spite of human nature, not in concert with it.

Studies of the living higher primates give us a clue of how our earliest ancestors lived. Within their social groups, essential to individual life and security, individual decisions were governed by shifting combinations of personal versus group interest. Groups were defined by cultural as well as biological criteria, especially various forms of reciprocal altruism and cultural group selection. Altruism within the group was strong, and matched by rivalry between groups.

Cultural selection is a process similar in principle to natural selection acting on genes but operating on memes (ideas). Humans inherit both memes and genes, and the study of their interactions defines the theory of gene-culture co-evolution. Conflicts between the two sorts of heritable information are common, and underlie most moral dilemmas, including those inherent in all forms of environmental activism.

Genes and culture may interact in various ways: culture may mediate natural selection, or genes may mediate cultural selection; or the two may enhance, ignore or oppose each other. Group selection at the cultural level which also mediates natural selection at the genetic level is possible but difficult to achieve.

Since arguments for environmental protection deal largely with moral issues, we need to understand the origins of morality. There are various conflicting models, all of which concede that morality is based on evolved behaviour patterns, although its full development is uniquely human. Acceptance of its evolutionary roots requires some

187

reassessment of traditional philosophical doctrines concerning freedom, equality and unconditional altruism.

Sociobiology, which applies evolutionary arguments to human behaviour (ignoring culture and group selection) seeks to understand morality in terms of gene-based selective advantage. To the extent that any moral behaviour advantages the group, say some theorists, it must work by self-deception. But this view can be criticised because it falls into the fallacy of misplaced concreteness, and because it confuses evolved, impersonal and unconscious biological altruism with cultural, personal, conscious morality.

Sociobiologists and other gene-centred theorists regard morality as a conscious rebellion against natural selection, and dismiss religion as merely another survival mechanism. But the Christian biologist can answer by pointing out that true human ethics might well be a later development of religion, just as evolution commonly takes a character evolved for one purpose and uses it for another. Morality can therefore be seen as the fulfilment of nature, not a rebellion against it.

It is more constructive to study the links and parallels between the natural, human and religious worlds without assuming the superiority of any one over the others. Each is able to contribute insights that illuminate the separately-developed understandings of the others. Their interactions define areas of conflict in the overlap zones, many of which are relevant to attempts to resolve the environmental crisis. Theologians can argue that there has been a progressive development from natural selection in nature, through cultural selection in human society, to the abandonment of all selection in true Christianity.

There is an important distinction between public goods, which are abundant enough to be available to every user regardless of the number of other users, and common-pool resources, which are limited in supply so that what is taken or damaged by one user is not available to other users.

There is also a spectrum of access rights, from private ownership at one end to free access at the other; in between is common property, to which access is restricted to a particular well defined group. When group decisions on management are well organised and free riders excluded, local commons can survive for centuries.

Most aspects of the human environment which are at risk (forests, local biodiversity, fisheries, clean water etc) are common property for which local management has degenerated into free access. The legal and social problems of organising just and equitable collective action to manage the common property of a very large group are complex.

If a system of social control of a common-pool resources breaks down, leading in effect to open access, the result is usually irreversible damage to the resource. Garett Hardin calls this the Tragedy of the Unmanaged Commons. It can be explained in terms of a Prisoner's Dilemma, in which each user plays against all the others as a group.

Three forms of the dilemma which can be observed in New Zealand are described. *Muldoon's Law*: In management of a common resource, strategies that are individually rational can be collectively disastrous. *Berk's Law*: The threat of damage to or depletion of an uncontrolled common resource increases its value and stimulates competition among free individuals to harvest it all the faster, regardless of the future. *Bolger's Law*: Individuals will tend to resist restriction of private access to common resources, even to protect the long term interests of the community.

To protect common-pool resources from over-exploitation, three main strategies are possible: privatisation, regulation and collective action. Each has its own advantages and disadvantages, depending on the type and location of the resource to be protected. All require monitoring with systems of rules and sanctions; the general absence of these explains why the declarations produced at Seoul and Rio have so far had little effect.

Promotion of church-based environmental activism to a sceptical world will inevitably invite public inspection of the doctrine of creation. Critical realism requires that we take the traditional view seriously but not literally. All credible theologies have to be continually updated in the light of contemporary secular knowledge, so this is not a new responsibility.

Science describes the processes of evolution under which the whole natural order developed from simple beginnings. It has a lot in common with the Old Testament creationist tradition, which asserts that God created the conditions under which natural processes were free to operate. Both allow for mistakes, which are the necessary corollary of freedom, and thereby explain the origin of evil. Both reject the Augustinian interpretation of the Eden myth and the commonly supposed need for nature to be redeemed. The integration of these two traditions is more a matter of elimination of errors and misunderstandings than of formulation of new doctrines.

The sociobiological theory of the origin of morality as self-deception is valid to the extent that many moral imperatives can be understood as good strategies giving a selective advantage in the evolved social game of tit-for-tat. But true Christian ethics goes beyond evolution, by deliberately abandoning all forms of social and natural selection and depending upon the grace of God.

Organised religion has an important role to play in the coming environmental crisis, provided it does not attempt to domesticate the debate by claiming that Christianity has been green all along, or to hang on to its old exclusivist claims to truth. It is one of very few organisations influential in our society with roots going back to before the modern crisis developed. It can take a cool but loving view of human nature and its multiple deviations.

Collective agreements on local environmental management need to take account of the stratification of most societies and the potential for disruption by the socially disadvantaged on the prospects of organising collective action.

Some Christians advocate reviving the Biblical idea of stewardship as a vehicle for the idea of caring for creation, but others point out that it is not easy to translate that ancient idea into the modern world: it assumes that humans are outside nature rather than part of it, that the reason God needs to be represented by a steward is that God is absent from the world, that it is possible to weight all competing interests, and that the differences between the various Biblical ideas about stewardship are unimportant.

We need to recover the relationist interpretation of the concept of *imago dei*, which would transform our concept of stewardship from dominion over nature to service to nature. This would require reconsideration of the modifications to Christian insights made in post-Constantinian times.

Religious involvement in the environmental debate requires far more than direct action, such as promoting recycling programmes, important though they are. It should concentrate the resources of the churches on attacking the causes of the environmental crisis rather than the symptoms of it. They should start in their own houses, by accelerating the already existing discussions on updating traditional creation theology in relation to secular knowledge.

The difficulties in the path of such revisions include the perceived threat to faith, the perceived distance between what goes on in theological academia and what the ordinary people believe, the general lack of scientific education among congregations, and the multiple consequences of tinkering with basic theological propositions. There is an urgent need to educate congregations on the metaphorical nature of theological truth, and to develop new metaphors that more clearly convey ancient truths to the modern mind.

The 1998 Hikoi of Hope demonstrated the potential for religious leadership in co-ordinating domestic and international protest against economic and environmental injustice. But religious leaders must also be prepared to consider the social position that the catholic church (Roman and Anglican versions) has enjoyed since Constantine, and the potential changes it might need to make in order to fulfil its primary prophetic role.

It is possible that the predicted environmental catastrophe will be averted, although no-one informed about the issues is optimistic, and time is much shorter than the general population believes. But if not, the churches still have a vital role to play in picking up the pieces. They have sources of strength and meaning supplied from outside the assumptions of free-market economics, and also are the only contemporary social organisations to have roots preceding two previous catastrophic social collapses (after the falls of Jerusalem and of the Roman empire). Therefore, churches that really live a vital, rational, passionate and compassionate form of Christianity will be a beacon of grace in a disintegrating society, and they have a solemn duty to prepare themselves for that role.

Appendix One:
How Natural Selection Works

The modern synthesis

The basis of our contemporary understanding of evolution by natural selection can be summarised in a few words: variation, competition, differential survival, and cumulative change (Table 4). It is called the modern synthesis because it is a twentieth-century integration of the two corner-stones of evolutionary theory, independently discovered by Charles Darwin and Gregor Mendel at roughly the same time in the nineteenth century. If Darwin had known of Mendel's proof of the particulate nature of inheritance as he developed his insights into the action of natural selection, the history of biology might have been very different. Further details are given by Williams (1992) and Maynard Smith and Szathmary (1995).

Table 4 How the processes of selection work in the animal and human worlds

Processes of selection	Animal world	Human world	Scripture
1. Variation in characters	Speed of horses	Makes of cars	Ex. 7:4: Israel vs Egypt
2. Units of information	Genes of successful sires	Copyright designs of components	Memorised texts
3. Handing on the heritable variations	Stud farms	Maker's style	Ex. 12: 26: Passover ceremony
4. Competition between traits	Races, shows	Free market	Deut. 30: 15: Moses' challenge
5.Best traits favoured	Bloodstock records	Consumer choice	Jos. 24: 15: Joshua's pledge
6. Accumulation of best traits	Wild pony to Sir Tristram	Horseless carriage to Rolls Royce	Mat. 5:17: Moses to Christ

The difference between information (genes) and materials (bodies)

It is well known that every living organism contains a unique collection of genes, heritable units of coded information. Together they co-operate to control the building of the body and all its unconscious functions such as breathing and digestion, and they influence important characters such as appearance and behaviour. But it is less well known that there is an important distinction between the genes and the individuals holding them. Grasping the significance of that distinction is the key to understanding modern evolutionary biology.

A famous champion racehorse sire such as Sir Tristram could not live indefinitely, but his owners' interests were not damaged by his death[1] because they have Zabeel, another young stallion of Sir Tristam's blood line, standing at stud in Cambridge, and at least eight others are doing the same for other owners. They, like all bloodstock breeders, work on the principle that, in the long term, genes are more important than any individual sire, because they will outlive him. Sir Tristram's body was the product of a unique, short-lived and unrepeatable combination of genes, compiled from the previous generation (his dam and sire) and split up in the next (his foals). By contrast, the genes themselves are effectively long-lived, since they replicate themselves with astonishing accuracy and pass on copies to each new generation of foals usually unchanged. The body called Sir Tristram eventually wore out, but the genes that made him a great racing sire will live on in his descendants and in the stud books.

Genes are carried by DNA, but it is the immaterial coded information, not the material molecule of DNA itself, which comprises the heritable data. The relationship between the genes and DNA is much like that between words and the paper they are written on: genes, like words, are units of information, whereas the DNA molecule, like paper, is only the material carrier of that information. As in human writing, sense is carried in the *order* of the three–letter units of code: CGA means something different from AGC, just as do DOG and GOD in English.

---

1. By the time he died in 1997 Sir Tristram was 25 years old and had fathered 44 Group One (the highest category) winners.

Genes replicate themselves from one generation to the next with astonishing fidelity: some genes controlling basic metabolic processes such as respiration and digestion are so stable, and mutations in them so immediately fatal and rapidly eliminated, that the copies we see have remained unchanged for millions of years (Williams 1992).

The genes will be mixed up in different sets in each foal, and not all the new combinations will be as successful as they were in Sir Tristram. In every generation, each new set of genes will build bodies that interact with each other, and not all bodies will be equally successful in passing copies of their genes to the next generation. In George Williams' terminology, individuals are physical, temporary, highly variable, disposable *interactors,* or bodies; genes, by contrast, are non-physical *codices,* replicable coded information, faithfully passed down through the generations until they are either altered by mutation or lost. Bodies interact with each other in the material domain, and genes interact in the information domain. Both are real, but only bodies can experience life. Sheet music depends on a real and intelligable code, but it is not real in the normal sense until a musician lifts it off the page and into our sensory experience.

Variation

Over the long term, new variations are added to the gene pool by chance mutations—imperfections in the processes of copying genes. Most such mutations are unhelpful, and are rapidly weeded out. A few of them are a genuine improvement over the alternatives on offer, and these are (rarely) incorporated into the pool of variation that accumulates over the generations. Over the short term, a second process, recombination, increases variation among individuals by ensuring that existing parental characters pass through to offspring in unexpected ways. The continual reshuffling of the variants available in the family gene pool produces, even among close relatives, many distinct individuals in every generation, all different in appearance and personality. The best stallions such as Sir Tristram serve only carefully selected and well-bred mares. Even so, many duds appear among the foals and are quickly sold off. This recombination process explains the common observation that, without ruthless selection, family dynasties, in racehorses or humans, never perpetuate the particular qualities of their founders into the indefinite future.

In all the higher animals, every individual has two copies of each gene, one inherited from each parent. The two copies usually form a complementary pair of alleles*, one dominant (usually expressed in the body) and one recessive (expressed only if there is no dominant allele of the same gene in the same body). They do not necessarily have the same effect. The two alleles are separated during meiotic* sex, the processes of producing eggs and sperm, so each germ cell carries only one copy of each; the pair are reunited at fertilisation. Very few physical conditions are controlled by only one or a few genes, but at least four thousand human disabilities are known to be caused by single-gene defects, such as cystic fibrosis or muscular dystrophy (Wolpert 1992: 166). Some are associated with the dominant of the two alleles, and appear in every child that inherits a copy; others are associated with the recessive allele, and usually remain unexpressed. Many people carry recessive alleles for serious diseases without ever being aware that they have them. They find out only if they happen to marry another unwitting carrier, in whom the effects of the recessive allele are also masked. The disease will show itself in every one of their offspring carrying two copies of the recessive allele.

Almost all other characteristics are controlled by a large number of genes, all in a great variety of combinations, and their effects are controlled by their interactions with each other and with still other moderator genes. Hence there is always an enormous amount of variation in the gene pool of any population, and in the potential variation in physical characters. The possible combinations are endless—far more than that in a pack of cards, which contains only fifty-two characters but can still be reshuffled and dealt out in a virtually unlimited range of possible hands. So the ancient Christian conviction that every individual human being is unique and irreplaceable is confirmed by genetics.

There is a great deal of duplication and redundancy in any species genome, which means that even though we can read the genetic code, we do not necessarily understand it. As Jones (1996) puts it, although the technology of ordering the letters in the DNA alphabet is well advanced, nothing is known about what most of it does. Working genes are, it seems, oases of sense in a desert of nonsense (xi).

Such extensive duplication allows room for some variations to remain hidden for long after they have outlived their usefulness. For example, the lower legs and single hooves of horses are in fact derived from the middle digit of the standard vertebrate five-digit foot. The genes controlling the other four digits are no longer useful, and are usually repressed. But they are still there, and in normal horses they produce the splint bones and the 'chestnuts'—a horny patch on the inside of each leg. Very occasionally an abnormal foal will turn up which has what look like two extra mini-hooves, because the full versions of the genes for two of the ancestral toes have for some reason shown themselves, inappropriately and out of context, in a modern animal that has no use for them. Animal breeders call three-toed horses and other such monsters 'throwbacks' and get rid of them immediately. Biologists, recognising their significance for the sciences of genetics and evolution, call them 'atavisms'—freak individual animals showing an ancient character—and put them in museums (Gould 1983a).

Natural selection and competition

The sheer fecundity of life is staggering, and easily demonstrated by any simple calculation based on the reproductive powers of rabbits or bacteria. But fecundity is not the same as reproductive success, and the general rule in nature is that there is nothing like enough room for all individuals or all variants to survive. Inevitably, some will live and others will die. The key question is, which?

Polar bears hunt seals on the winter sea-ice of the north polar basin, but that habitat has been available only since the Pleistocene period, about two million years ago. The common ancestors of contemporary polar and grizzly bears were certainly brown or black, and those that first explored this new habitat were easily spotted against the snowy background. In the competition for survival, bears that happened to develop a gene[2] for white fur were more likely than others to survive and breed, and pass on the genes for white fur to their young. It is probably safe to predict that the gene for white fur is so successful that every single living polar bear must have a copy of it, even though the mutation is so recent (polar bears have been a distinct species for <100,000 years) that they retain the dark skin of their brown-furred ancestors underneath their white fur (Jones 1996: 185). Such successful genes are "fit"; ie, they have selective value appropriate to the conditions. Conversely, natural selection eliminates unfit genes from the gene pool—probably very rapidly in the case of any reappearance in polar bears of the gene for their ancestral brown fur.

Animals face competition both within and between kinds. Hyenas have to compete both with lions for game and with each other for meat at the kill; the European rats that colonised New Zealand from the settler's ships competed for living space both with each other and with the Polynesian rats that had arrived with the ancestors of the Maori. Wherever there is variation and competition, there will be *differential* survival. Natural selection is not the same as a disaster, such as a landslide which kills everything in its path, good and bad alike. It discriminates between variants, favouring some and eliminating others. There is nothing deliberate about it: it is just that individual animals carry different combinations of genes sampled from the common gene pool of the population, and some will be more successful than others in returning copies of their genes back to the pool (Williams 1992). The basic rule for success in genetic competition is: 'Look after Number One' (Table 4)—it cannot work in any other way.

Variation is the essential pre-requisite for natural selection, and, over time, for evolution. The ultimate sources of variation (mutation and, more often, recombination) are random in origin, but that does not make them disorderly: in physics also, orderly macro-level phenomena are derived from random micro-level events (Bartholomew 1984). Although it was chance that produced the original allele(s) for white fur in the early polar bears, the reward for the individuals that carried them in a snowy environment was entirely predictable from the rules of natural selection. It is the *interplay* of chance and law that makes evolution both entirely logical and at the same time also entirely

---

2. The technically correct but unwieldy statement here should be 'bears in which a mutation at the locus of the gene for fur colour provided an allele for white', but the normal, simplified terminology is easier reading.

unpredictable (Peacocke, 1993). The role of chance in long term evolutionary change is a huge subject in itself, but over the short term it is less important than the small changes in gene frequencies from one generation to the next, which are caused by the orderly processes of natural selection. The cumulative result is long-term genetic adaptation.

Adaptation: the long-term, accumulative process of change

There are a very few extremely stable environments, such as the deep sea, in which ancient forms can survive virtually unchanged for millions of years—such as the so-called 'living fossil' fish, the coelocanth. But over most of the earth's surface the natural environment is always gradually changing, on time scales ranging from daily to glacial. If the changes are very slow, existing species can keep up with them by 'Red Queen' processes, named after the character in *Alice in Wonderland* who had to keep on running in order to stay in the same place. If the long-term result of continuous adaptive change in the same direction is a new species fitted for life in a new environment, it can lead to speciation by descent. One of the best examples of this comes from New Zealand. About eight to ten million years ago the South Island was virtually all flat land covered in lowland forest and inhabited by parrots much like modern kaka. As the earth shook and the Southern Alps slowly built up under their feet, the birds adapted gene by gene as the altitude of their homeland increased inch by inch. These slow, cumulative changes in the landscape and in the birds eventually produced (among other alpine forms), the tussock grasslands and the kea (Fleming 1962).

Where such processes have been left undisturbed for a long time, and local populations become separated by barriers of habit or geography for a few aeons, the result can be a great range of new species. For example, a large, stable, ancient continental landmass offers a huge variety of habitats for birds that feed in, say, the kingfisher fashion, and over the millenia the different local breeding groups have accumulated slightly different sets of genetic changes causing local variation in colour, size or behaviour. If two non-interbreeding groups remain

199

separate for long enough, they will eventually develop into separate species—mutually infertile by definition—and if they meet again they can co-exist without losing their separate identities. New Zealand, a recent and much smaller landmass than any continent, has only one species of kingfisher (and that is a recent immigrant), whereas North America has three, Australia has eight, and South Africa has ten.

By contrast, temporary and local variation can be met by small, often reversible adjustments. Where local variants meet, they may hybridise into a still another form, if the differences between them have not reached the level of full species. Red deer and wapiti deer have done that in New Zealand, the only country in the world that supports both. The escaped domestic pigs of the early New Zealand settlers, which within a few generations reverted to the fierce and hairy boar-like appearance of their original ancestors, are not a new species because they would still be able to interbreed with true wild boars if there were any here.

In sum, adaptation is the continuing running adjustment of living things to their habitats. The ecologist GE Hutchinson (1965) compares the process with an 'ecological play' producing new characters during one long continuous performance in an "evolutionary theatre".

Natural selection is a very slow process, because by definition it works by differential survival and *reproductive success* of individuals. But over the very long term—tens of millions of years—the mechanism of natural selection applied to tiny increments of advantage can produce unimaginable changes. For example, the earliest whales were descended from terrestrial animals that returned to the sea complete with ordinary, doglike paws. The long-term accumulation of the complex of genes controlling small differences between individual early whale-ancestors in paw width eventually gave their descendents broad swimming flippers, still based on the bone structure of the five-toed paw of their predecessors.

The agents of natural selection can be anything—shortage of food, severe climate, predators, parasites—that challenges individual animals to show what they are made of. Stress serves to distinguish between those that happen to be carrying appropriate genes and those that are not. Natural selection is nature's equivalent of a kennel club judge asking the competitors to show off their paces. The process used to be called 'the survival of the fittest', but that is a misleading name, since it is not the animals themselves that are being judged fit or not, but their genes. In the same way, ultimately, kennel club judges concentrate on the characters that meet the breed specifications rather than the individual animals in which they happen to be represented this year. The current champion dog won the crown because it demonstrates those breed specifications more perfectly than any other dog, not because it is *itself* perfect.

Natural selection is not itself a creative process—it does not *create* new and fit variations, only eliminates unfit ones. But it frequently produces new structures by adaptation of existing ones. For example, the small and specialised bones that form the mammalian middle ear and voice box can be traced back over the millenia to gill arches of far distant ancestral fish (Gould 1993a). The air sacs that enabled certain primitive fish to survive temporary droughts were modified by their descendants, the modern bony fish, into the swim bladder that provides neutral buoyancy in water of any depth and pressure (Gould 1993c). The standard vertebrate limb has produced the otter's flipper, the cat's paw, the horse's hoof and—by different and independent adjustments—the wings of both bats and birds.

Because competition so often eliminates the weaker of two species that are too much alike, it is usually safe to infer that similar types of animals or plants which still survive have developed alternative ways of life that allow them to live together *without* competing. For example, there are plants that specialise in colonising bare ground, and plants that come after them in an orderly succession from low, quick-growing, short-lived shrubs to the tall, slow-growing, long-lived forest trees. Within many long-established forests, each of the dozen different kinds of birds that can be found together in any given area forage in the trees in a slightly different way. They do not compete because they are searching for food in slightly different places.

Evolution is the unplanned *consequence* of all these processes—variation, competition, natural selection, differential survival and cumulative changes in gene frequencies—generating and then choosing between more or less appropriate sets of characters that decide which individuals will succeed in surviving and breeding—thereby returning those genes to the species gene pool. Natural selection operates to eliminate the unfit *individual* bodies plus the genes they carry, and to accumulate the fittest genes from the bodies that best survived and bred in the previous generation. The genes that control the best characters produce the only bodies we see; all the rest have been eliminated. As S J Gould says, natural selection is not a creative process; it is a sieve, not a sculptor (Gould 1993b: 317)—ie, it cannot create fit organisms, only eliminate the unfit.

The major transitions of evolution

Modern living organisms are unbelievably complex, but they are derived from much simpler ancestors over the (roughly) three thousand million years since life began. That implies an increase in complexity over time, which is true in some lineages but not in all, and was not inevitable in any. Maynard Smith and Szathmary (1995) have assembled an impressive account of the history of life, showing that the increase in complexity has depended on a series of significant transitions in the way in which genetic information is transmitted between generations.

Of the eight transitions they list, five took place before the beginning of the fossil record, so deductions about them were impossible before the development of molecular biology. Some were unique, such as the origins of the eukaryotic* cell, of meiotic* sex and of the genetic code itself. Others, such as the origins of multicellular organisms and of animal societies, have happened several times independently.

After every transition from one level of organisation to the next, entities that were capable of independent replication before the transition could replicate only as part of a larger whole after it. This observation raises a problem common to each transition in turn: why did the formerly independent entities allow themselves to be taken over? Why did not natural selection choosing between entities at each lower level prevent any integration at the next higher one?

The answer most consistent with the facts is that of immediate selective advantage to individual genes. For example, the cells of all multicellular organisms are descended from single-celled protists, once capable of surviving on their own, but today they can exist only as parts of a larger organism. But all multicellular organisms develop from a single fertilised egg, so all their cells are genetically identical, and the genes carried by each cell can survive only if they all co-operate to build a functional multicellular body. It is only where the genetic interests of all parties are identical[3] that there is no competition. This insight from cell biology is enormously significant in understanding failures in human co-operation, from Brennan's question (p.71) to the Body of Christ.

Two processes, symbiosis* and epigenesis*, are particularly significant here. *Symbiosis* means the living together of entities that are not genetically identical. Because their interests are not the same, there is a range of possible outcomes, from mutualism beneficial to both, to outright exploitation of one by the other (Table 1). The long history of life has produced a huge variety of animals and plants that have co-evolved to help and exploit each other in complex ways. Because related individuals have more interests in common, the degree of their co-operation is predictable from the degree of their relatedness. In animals as in human life, individuals are more likely to put themselves out for their offspring or their siblings than for their second cousins. *Epigenesis* refers to an increase in complexity which is not due to genetic change in the usual sense (at the level of the DNA), but to the fact that different genes are active in different cells (so muscles, bones, nerves and blood all develop from an identical set of instructions) or in different castes of the social insects (workers, soldiers, drones)—and, perhaps, in different people.

Social behaviour in humans and other mammals is especially interesting because it involves close co-operation between individuals that are all genetically different. To understand how sociality arose, we must explain why independent individuals live together and tolerate all the various problems of communal life—including conflicts of interest at various levels and, often, reproductive disadvantages. To understand the maintenance of animal societies (including human ones) we must

---

3. Later in life, accumulating mutations in the somatic cells introduce relentlessly increasing genetic differences within the body, which sometimes allow cancer and always lead to ageing, and death (Jones 1996).

explain how they manage to resist cheating (Maynard Smith and Szathmary 1995: 264). These are old questions—they have puzzled thoughtful people since Biblical times—but answers have been developed only in the last thirty years. In due course we hope that these answers will help us to understand why any form of environmental protection which depends on getting everyone to co-operate for the public good is usually so difficult.

Conflicts of interest and the levels of selection

Until the 1960s it was commonly assumed that natural selection in the non-human world must always work for the good of the species. Christians of the Teilhardian school liked this interpretation, because it readily fitted with their idea of the workings of God pushing the collective development of creation in the direction of the foreordained emergence of humans capable of making moral decisions (Galleni 1992). The problem is that this view cannot account for many contrary observations in the real world, which are inescapably real but very *bad* for the species. The keys to understanding them are to recognise (1) the difference between the realm of short-lived, material bodies and the realm of long-lived, immaterial information, and (2) the hierarchical organisation of nature and the multiple levels (within-group and between-group) of selection (Wilson 1997b).

Like the kennel club judge, natural selection chooses primarily between alternative sets of genes, ie between particular heritable characters. It has to select between bodies interacting with each other in the courses of their lives, because genes can only show their effects in bodies. But, *un*like the kennel club judge, natural selection is totally uninterested in the long-term good of the breed. So male lions taking over a pride of females soon kill all the existing cubs in order to reduce the time before the females will come into heat and produce cubs for their new mates. It seems cruel and wasteful for the species as a whole, but that is irrelevant to the outcome. What matters in the long term is that the behaviour pattern that survives most often is the one in which the next generation of cubs is fathered by the incoming males. Different behaviour patterns can be explained in the same terms. For example, male foxes taking over a new mate do not kill the cubs sired by their mate's previous partner, but that is not because foxes are more compassionate than lions. The reproductive cycle of the vixen is seasonal, and she could not immediately produce cubs for her new partner whether the previous litter survived or not. Meanwhile, if the existing cubs grow up but cannot find territories of their own, one or more may stay and help their mother and stepfather rear the next generation of cubs. The explanations for why such variations in parental attitudes among male animals persist became clear during a decisive demolition of the good-of-the-species explanation by Williams (1966).

Current synthesis theory asserts that only genes (or their alleles) which control behaviours that have some advantage over the alternatives available in this generation will be copied and passed on to the next generation. The genes of male lions of type A that kill their predecessor's cubs will be passed sooner to a new generation of cubs than the genes of hypothetical males of type B that spare their rival's type A existing cubs 'for the good of the species'. Within a few generations, Type A genes will predominate, and the more merciful type B genes will be eliminated from the population. Type A genes control behaviour that has the effect of advancing their own interests, whereas Type B genes do not. Dawkins (1989) used the metaphor *The Selfish Gene* to explain this process to general readers. Theoretical models based on the selective advantages of particular genes have been spectacularly successful in stimulating research and explaining a huge range of observations in animals, and they have generated an enormous literature and very many important insights into biological processes. They are also a necessary, but *not* a sufficient, pre-requisite for understanding the biological basis of human behaviour (p.79), plus some unexpected insights into biblical events (Table 5).

Altruism: the good of the species?

Tennyson's description of nature as 'red in tooth and claw' is well-known but not entirely correct. The natural world also provides many examples of what looks like real altruism. A mother bear defending her cubs is a truly dangerous animal, ready to take real risks to her own life

205

in order to drive away any threat to theirs. Adolescent jackals and foxes regularly stay at home to help their parents rear the next litter of cubs, and will feed and play with the young as if they were their own—even to the extent of taking over complete responsibility and successfully rearing the litter if the parents are killed. All the lionesses of a pride tend to produce their cubs at about the same time and suckle them all indiscriminately, so that a cub can get milk from its aunts or adult cousins as well as its mother. Every worker bee in a swarm is a non-breeding female which will never produce offspring of her own but helps her mother, the queen, produce more workers. Oxpeckers pick ticks off antelopes, cleaner fish spruce up the gills of larger fish, mongooses and hornbills forage together, the mongooses stirring up insects for the hornbills and the hornbills warning the mongooses of approaching enemies.

Darwin knew that his theory of natural selection should make altruism in nature impossible—certainly between species:

> As in nature selection can act only through the good of the individual, including both sexes, the young, & in social animals the community, no modification can be effected in it for the advantage of other species; & if in any organism [such] could be shown to exist . . . it would be fatal to our theory (Stauffer 1975).

His dilemma arose because he could see clearly the conflict between the ruthless logical consequences of his own theory and the apparently altruistic behaviour of many real animals. He proposed the solution, at least for the development of altruism *within* species, that natural selection can choose between *sets* (he calls them 'communities' in the quote above) of bodies carrying alternative characters. For this example, that would mean asking if populations of lions in which the new males adopted the existing cubs for the good of the species, in order to avoid wasting the food and care already invested in them, could do better than populations of lions that did not. The answer is, they cannot.

Darwin's acceptance of group selection went unremarked until the 1960s. But for the last thirty years the idea has been rejected, largely because it requires natural selection to make choices between related but separate groups of individuals of the same species. But in nature, interbreeding too easily destroys the differences between related groups, and within each group, individual self-interest is nearly always stronger and more effective than group interest. Only if all members of a group share the same fate—like the genes within a body, or the rowers within a crew—does selection operate at group level (Wilson 1997b). Genetic selection at the group level in nature usually does not work because nature so rarely provides the conditions for competition between genetically defined groups (within or between species)—the natural equivalents of the football league.

Altruism between relatives: kin selection

Opponents of the evolutionary interpretation of nature like to point to examples of 'natural' altruism, such as the co-operative lionesses and jackals, as proofs that science is quite wrong—that nature was, as Genesis asserts, created good as well as beautiful, and it was only human greed and disobedience that has spoiled it.

On the contrary, these examples of animal altruism within species confirm the modern synthesis: they are all cases of kin selection, which still obeys the universal principle, 'look after number one'. The difference is that here, the 'number one' being looked after is the gene, not the body it happens to be in for this generation. It works because of the crucial difference between the genes and the bodies they inhabit.

Kin selection is a simple modification of natural selection. It relies on the initially counter-intuitive fact that, in the calculus of evolution, reproductive success can be as well be achieved by proxy as by producing young of one's own. If animals are related, they share copies of the genes they inherited from their common parents. Relatedness reduces the extent of competition and increases the chances of co-operation. In mammals and birds, parents and their offspring share fifty per cent of their genes, and so do full siblings. Any altruistic act by one individual that benefits a related one will increase the chances of some copies of the genes they share reaching the next generation, even if the altruistic individual is disadvantaged personally. Since all the lionesses in a pride are related to each other, all the cubs carry copies of the females' genes, and it scarcely matters who feeds which. Maiden aunts who help their sisters rear young at the cost of their own family prospects are still helping to pass on copies of the genes they all carry. This apparently altruistic action contributes to the 'inclusive fitness' of the aunt, and can be favoured by natural selection. Likewise, the young jackals that babysit their infant siblings are doing more to further the interests of their own genes than if they left home. Jackal cubs reared by several adults have far more chance of surviving than those reared by solo pairs, and all carry copies of the altruists' genes.

The most spectacular examples of kin selection are the Hymenoptera (the sociable bees and ants), but that is because they have a peculiar genetic system, completely unlike that of humans. Worker bees share *more* genes with their siblings than they would with their own young, so their genes would be positively penalised if all the workers decided to go it alone. Despite the ancient comment 'go to the ant, thou sluggard', the industry and self-sacrifice of ants and worker bees should not be held up as an example to people, since the mathematics of relationships argue against the emulation of their unconscious altruism being developed on the same scale by the members of any human society.

If the theory of kin selection is true, the degree of co-operation between individuals in all species capable of recognising who their relatives are should depend directly on the degree of relatedness. The highest relatedness possible, in all species except the hymenopterans, is the fifty per cent between a parent and its offspring and between siblings. The theory therefore predicts that the greatest degree of altruism is expected between these closest relatives. That is in fact what is found, and it explains many oddities in nature. Why is parental care observed much more often in females than in males? Because in animals parental care is a good strategy only to the degree that the parent can be sure the offspring is its own (internal fertilisation and long gestation mean that, at least in mammals, maternity is a matter of fact, paternity is a matter of opinion). Why does a male lion have a better chance of joining a pride if it has a brother? Because the coalitions between males that are often necessary to take over a pride are best formed by brothers. This process of favouring relatives at the apparent (but not real) cost to one's own genetic inheritance, so far from disproving the neo-Darwinian view of nature, has proved to be one of the most spectacularly successful explanations of a wide range of hitherto inexplicable observations.

The theory of kin selection also allows remarkably detailed analyses of parental investment among animals. The mother bear who defends her cubs against a marauding puma is in fact serving the purposes of her genes and the copies of them that the cubs already carry. On the other hand, if her altruistic action costs her her life, she is giving up the chance of producing more cubs in the future. Each litter of cubs demands an investment of time and energy by the mother, which reduces her chances of producing another litter in the future. Mother animals tend to invest just as much in each batch of young as they need to have a fighting chance of survival, but not more, since any surplus maternal effort would be better saved for a future litter. So when the young of one litter are old enough to fend for themselves, the mother bear evicts them and starts on the next lot. Since every cub shares half the mother's genes, they share some of her interests and usually go after only minimal resistance, although the half of their genes that are not shared with the mother favour the cub's efforts to persuade her to let it stay a bit longer. The interplay between co-operation and competition places limits to altruism in nature, which can be calculated from the degree of relationship and the life expectancy of the particular interacting individuals. The point of balance is achieved in retrospect and quite unconsciously by natural selection, in its constant work of distinguishing fit from unfit genes.

Sibling chicks of many birds, such as raptors and pelicans, are graded by size; the older ones generally gang up on the smallest, which seldom survives unless food is very abundant.[4] Newborn hyenas come into the world equipped with eyes already open, teeth already erupted and awesome tempers; they fight each other almost as soon as they are out of the placental membranes. Paradoxically, the explanation for disharmony between siblings of some species is the same as that for the helpful nature of the siblings of other species such as jackals: siblings share half each other's genes. The half they have in common favours the

---

4. This observation causes anguish to theologians (McDaniel 1989) but is not incompatible with authentic Christian creation theology (p.125).

tendency to help each other; the half that is different favours ruthless competition, especially for the food and other resources provided by the mother. Which tendency dominates depends on the evolutionary history of the species and the circumstances of the time, especially the distribution, size and supply of food.

Altruism between unrelated individuals: reciprocity

Not all altruism is driven by kinship, in either the animal or the human worlds. Kin selection is certainly a powerful basis for apparently unselfish behaviour in family groups, but reciprocal altruism provides another mechanism that operates in groups of unrelated individuals. The main difference is that the predictability of the expected benefit depends not on relatedness, as in kin selection, but on the probability of reciprocation. In turn that depends on a good memory, permanence of association and other characters unaffected by relatedness.

It is particularly clear in intelligent animals that have good memories and extensive social relationships, such as the social carnivores (Macdonald 1983). In meerkats and dwarf mongooses, the adult members of a pack are not necessarily related, but they take turns to look out for predators while the rest of the pack feeds. As all the infants in the group are the offspring of only the dominant pair, many adults have to put their own parental aspirations on hold, and some never get the chance to breed at all. But the subordinates cannot survive long outside a pack in which mutual defence against predators gives an individual enough time to forage for itself, so they tolerate subordination within a pack as being a better strategy than the likely alternative, death outside one. They will be well placed to compete for the top spot when it next falls vacant, and in the meantime they work diligently to gain valuable experience in parenting skills.

Reciprocity extends altruism outside the family. It allows a cohesive social structure in groups including non-relatives, provided each individual is able to recognise other individuals and remember past favours. It works so long as A, having benefited from B's help today, is willing to help B tomorrow. In a small group where all individuals know each other and can remember past interactions—that is, in many of the higher mammals and all pre-humans and humans—the problem is to avoid exploitation by cheats who gain now but refuse to reciprocate later. B must remember A's debt and retaliate if A refuses to pay it, by refusing further co-operation or by active aggression. For example, the benefit conveyed by one chimp that grooms another troubled with parasites easily exceeds the cost of the action. Therefore, grooming alliances in primates are easily established, and are very important means of reinforcing the bonds of friendship between certain individuals, as well as the rules of reciprocity among all members of the group. But cheaters become well known, and have difficulty in finding a grooming partner when they need one. Social rejection of cheaters is one of the most powerful means by which sociable animals maintain systems of reciprocal altruism, and the extension of that mechanism into human life underpins many of our social attitudes, such as disapproval of those who attempt to 'free-ride' on buses, or fail to return dinner invitations or write thank-you letters (Table 5).

Table 5 The continuity of selection processes in the animal, human and religious worlds, and Jesus' response showing the consequences of God's abandonment of all selection

Mode of selection	Proverbial description	Natural world	Secular world	Religious world	Jesus' response
Individual selection	"Look after Number One"	Flowers	Fashion/fast cars	Solomon (1 Ki 10:4)	Mat 6:29
		Birdsong	Army displays	Goliath (1 Sam 16:4,11)	Mat 26:52
		Rats at tip	Wealth	Rich fool (Lk 12:19)	Lk 12:21
		Raptor chicks	Sibling rivalry	Cain/Abel (Gen 4:4-5)	Mat 5:22
		Cuckoos	Adultery law	Stoning (Jn 8:4-5)	John 8:11
		Pack hierarchy	Competition for social status	Hypocrites (Mat 6:5b) Who is the greatest? (Mk 9:33-4)	Mat 6:5a Mat 9:35
		Parasites	Bludgers	No work, no eat (2Th 2:10)	Jn 13: 34
Kin selection (relatives)	"Blood is thicker than water"	Lionesses	Extended families	Rich man's brother. (Lk 16:27)	Mk 3:33-5
		Bear & cubs	Any mother	Solomon's judgement (1Ki3:27)	Mt 10:37
Reciprocal altruism (non-relatives)	"You scratch my back, I'll scratch yours"	Mongooses	Treaties	Solomon (1 Ki 3:1)	Mat 5: 46-7
		Ox peckers	Mutual dinner invites	Golden rule (Mt 7:12)	Lk 14: 13-14
Group selection	"All for one, one for all"	Meerkats	Tribal societies	Moses' loyalty (Heb 11:24-6)	Mt 5:43-4
Beyond selection	Unconditional love	No equivalent	Romantic idealism	Abraham (Gen 22:12)	Mt 10: 38-9 Jn 15:12-15

Altruism: multi-level and group selection

Animal bodies are composed of subunits, including genes, and are themselves subunits of larger entities such as social groups. A recent extension of natural selection theory sees it as a multi-level process that operates on a nested hierarchy of units (Wilson 1997b). When all the members of a group have the same fitness and a shared fate (like all the organs of one body, or the eight members of a rowing crew), they lose their individuality and are treated by natural selection as subunits of a larger whole. Because selection can choose only between whole units differing in fitness, it cannot distinguish between the separate organs of a body or the separate rowers of a crew, because they are similar sub-units of a larger unit. It has to shift up a level until it can choose between separate bodies, or boats, that have different fitness and are in direct competition.

In a normal body all the cells contain identical copies of the same set of genes present on the fertilized egg—at that stage, the set in the egg was the *only* set that existed. But mutations may arise in some genes during the repeated copying of that original set through the life of the individual animal, and those altered genes may differ in fitness from the rest. Then they become competing sub-units, and the individual body is their environment. 'Outlaw' or 'rogue' genes occasionally appear, which have effects favouring their own propagation at the expense of the integration of the whole body (Dugatkin and others, 1994). They are normally counteracted by 'the parliament of genes' which suppresses the outlaws and enforces the co-adaptation of all the units of the genome. The same applies to somatic cells: cancer develops when mutated normal cells escape the control over growth and differentiation exerted by the genes, and begin to multiply at the expense of the rest of the body.

At the other end of the scale, in a colony of ants, each individual ant is an organ, a subunit, of a genetically defined organism, a colony, which is in competition with other such organisms. This arrangement means therefore, the social insects are among the very few examples in nature of genetic selection at the group level. In humans, altruism at the group level is very important, but controlled largely by cultural, not genetic, mechanisms (p.79).

Appendix Two: Some Organisations Supplying Relevant Information

UK and Europe

Christians in Science: 88 Sylvandale, Welwyn Garden City, Herts AL7 2HT.

The Institute for Contemporary Christianity: Christian Impact, St Peter's Church, Vere St, London W1M 9HP.

The John Ray Initiative: 5 Chancery Lane, Clifford's Inn, London EC4 1BU.

The Religious Education and Environment Programme: Rodwell House, Middlesex St, London E1 7HJ.

Real World Coalition: 17 Carlton House Terrace, London SW1

Science and Religion Forum: St Alban's Vicarage, Mercer Ave, Coventry CV2 4PQ.

World Council of Churches, Programme Unit III - Justice, Peace and Creation: PO Box 2100, 1211 Geneva 2, Switzerland.

USA

Center for Theology and the Natural Sciences: 2400 Ridge Road, Berkeley CA 94709.

Institute for Religion in an Age of Science: PO Box 341, Quakertown NJ 08868-0341.

National Religious Partnership for the Environment: 1047 Amsterdam Ave, New York 10025 (www.npre.org).

Science and Spirit Resources Inc: 171 Rumford St, Concord NH 03301-4579.

United Nations Environment Programme: DC2-0803 United Nations, New York NY 10017.

Glossary

Allele:

One of a pair or more of genes at the same locus on a given chromosome.

Atavism:

An ancestral character not usually observed in modern animals.

Church:

Used with a capital letter in the present context, this means the Anglican Church, especially the Church of England and the Church of the Province of New Zealand. This usage is not a value judgement on any other church, but merely an acknowledgement that it was St John's College Trust (Auckland) that funded my research and has first call on the results, and that I did much of the writing while on sabbatical in Oxford. Used without a capital, it means the Christian church in general. Full ecumenical relevance is assumed throughout, not only to other Christian denominations but also to the major non-Christian faiths as well. As Hall (1986:41) put it, 'there can be no responsible theology now that is not global in its perspective'. However, ecumenism does not require heterodoxy or uncontrolled eclectism, an unfortunate outcome of the Assisi Declarations (Berry 1995).

Critical realism:

The view that reality precedes theorising, which in turn implies that the real world must place distinct limits on theological speculation.

Default setting:

In a computer, the normal setting of all

programme options, the basic start-up arrangement to which it always returns unless over-ruled. Used here as a metaphor for 'basic' human nature.

Deism:　The concept of a distant and impersonal God as the Creator who made the world but then took no further interest in its workings, in contradistinction to theism, the concept of a transcendent and personal God as both Creator and constant sustainer of the world, passionately involved in all its workings.

Diploid:　The condition of having paired chromosomes.

EFM:　Education for Ministry, a four-year theological study programme for lay people.

Ethology:　Study of the evolution and adaptive functions of animal behaviour.

Epigenesis:　Theory that the embryo is formed by successive changes in structure, and differentation of, the action of genes in different cells.

Eukaryotic cell:　Cells whose DNA is organised into chromosomes with a protein coat and surrounded by a nuclear membrane.

Evolutionary egoism:　Self-interest at the genetic level, not involving personal or morally accountable choices by an individual.

Evolutionary psychology:　A more recent and less political version of sociobiology (qv)

Fallacy of misplaced concreteness:　The tendency to organise knowledge in terms of abstractions and then to reach

215

conclusions and apply them to the real world as if abstractions and reality are the same thing.

Hierarchy: A system of rank order among social animals, including humans. Also, metaphorically, the system of levels of organisation of the sciences (see *reductionism*).

Incarnation: The doctrine affirming that the divine nature of God assumed human nature in Christ.

Integrity of creation: See variety of definitions discussed on pp.132-137.

IRAS: Institute for Religion in an Age of Science.

JPIC: The Justice, Peace and Integrity of Creation programme (JPIC) of the World Council of Churches (WCC), now called Unit III, Theology of Life, based in Geneva.

Kin selection: A theory explaining the evolution of altruism among animals. The actions of one animal that benefit the breeding success of its relatives at the cost of its own can effectively transmit to the next generation copies of the genes they have in common, even if the altruist does not itself breed. It predicts that the degree of co-operation between individuals, in all species capable of recognising who their relatives are, will depend directly on the degree of their relatedness.

Macrophase wisdom: Term used by Brian Swimme to indicate the potential human ability, much-needed in the modern world but not nearly common enough yet, to make decisions transcending the evolved imperatives of our ancestral tribal groups. See Microphase wisdom

Maximum
sustainable yield: The cropping rate that will produce the largest possible harvest of individuals without reducing the standing crop or the future productivity of the population.

Meiosis
(Reduction division): A form of cell division, found only in the sex cells, in which copies of the paired (diploid) parental chromosomes are split into single strands and separated in the daughter cells. See Meiosis

Meiotic sex: Sexual reproduction involving meiosis, in which each partner produces gametes (eggs or sperm) each carrying only a single chromosome (ie, half the genetic information carries by each parent). At fertilisation, the single-strand chromosomes pair up again, restoring the diploid condition. The process of pairing up allows mutual cross-checking for copying errors.

Metapopulation: An aggregate of several separate interbreeding populations which together hold the species total or local gene pool.

Microphase wisdom: Term used by Brian Swimme for the evolved minds and attitudes of our early ancestors, adapted to tribal-level rather than global-level politics.

Multi-level
selection theory: Recognition that natural selection works on a nested hierarchy of units, from genes through organisms to groups. Selection distinguishes between sets of units that compete with other sets, but does not distinguish between units within a set that all share the same fate.

Myth: A narrative elaboration of culturally shared perceptions of reality, historically always the template for the derivation of moral values (Rue, 1989).

Naturalistic fallacy: GE Moore's term for the attempt to define goodness from natural properties (Honderich 1995:606).

Non-zero-sum games: See zero-sum games.

North: Current euphemism for the 'developed' world regardless of location, including New Zealand. By analogy, the term may be extended to individuals among the Southern social elites who have adopted the pursuit of growth economics.

Reciprocity: A process of holding social groups together by reciprocal altruism, or the trading of favours.

Reductionism: The natural world is a hierarchy of levels of organisation, increasing in complexity from physics through chemistry to biology. Reductionism aims to understand each level by breaking it down into component parts that can be analysed in terms of the level below it. 'Hard-line' reductionists claim that all biology can be explained in terms of chemistry, and ultimately, physics; no other explanations are 'real'. Their opponents point out that many features of the higher levels of organisation are non-reducible emergent characters. They cannot be understood at all except in terms of concepts appropriate to that level, and there is no reason to suppose atoms are more 'real' than a person or a social fact (Peacocke 1993: 39-41).

Religion: See p.13.

Science: See p.9.

Selection: Differential survival of self-replicating biological or cultural units between one generation and the next. *Kin selection* discriminates between genes, *natural selection* between individuals, and *group selection* between distinct populations. All three apply to humans, but only the first two to animals and plants. See Table 2.

Sociobiology: Study of the evolutionary background and adaptive functions of social behaviour in animals and humans.

South: Current euphemism for the 'under-developed' world regardless of location, including certain groups within the North who are disadvantaged by growth economics. Preferred by some writers to the older term 'Third World', which, in terms of numbers of people, would more accurately be called the 'Two-thirds World'.

Sustainable development: Development which meets the needs of the present without compromising the ability of future generations to meet their own needs (Bruntland 1987).

Symbiosis: Co-habitation of entities that are not genetically identical.

Theism: See Deism.

Theology: As defined by (Ambler 1990): serious intellectual enquiry into matters of spiritual concern.

UNCED: United Nations Conference on Environment and Development, held at Rio de Janiero in June 1992, also known as the Earth Summit.

Vernacular egoism: Personal selfishness, the consequences of conscious, morally accountable choices made by the individual.

Zero-sum game: A game in which one side must win and the other side must lose, such as almost all high-profile competitive sports. Contrast with *Non-zero-sum* games, systems of co-operation in which *both* sides win, and the more they co-operate the greater the rewards, such as most social interactions.

References

Alexander R D. 1987. The Biology of Moral Systems *(New York: Aldine de Gruyter),* 301 p.

Ambler R. 1990. Global Theology: The meaning of faith in the present world crisis *(London: SCM Press)* 90 p.

Anon. 1990a. Justice, Peace and the Integrity of Creation in Aotearoa New Zealand *(Auckland: Conference of Churches in Aotearoa New Zealand, PO Box 9573, Newmarket AK),* 48 p.

Anon. 1990b. Mission in a broken world: Report of ACC-8 Wales 1990 *(London: Anglican Consultative Council),* 186 p.

Anon, editor. 1994. Putting Creation in its Place *(Palmerston North, 26 November 1994: Diocese of Wellington),* 67 p.

Anon. 1995. 'Religions vow a new alliance for conservation' *One Country 7 (April-June 1995):1,* 12-14.

Anon. 1998. Land use control under the Resource Management Act; analysis of submissions *(Wellington: NZ Ministry for the Environment, Wellington),* 65 p.

Appleyard B. 1992. Understanding the Present: Science and the Soul of Modern Man *(London: Picador),* 283 p.

Arnhart L. 1998. 'The search for a Darwinian science of ethics' *Science and Spirit 9(1):4-7.*

Athanasiou T. 1996. Slow Reckoning: The Ecology of a Divided Planet *(London: Secker and Warburg),* 385 p.

Attfield R. 1983. The Ethics of Enviromental Concern *(Oxford: Basil Blackwell),* 220 p.

Attfield R, Dell K. editors. 1996. Values, Conflict and the Enviroment. Second ed. *(Aldershot: Avebury Press),* 174 p.

Austin W. 1980. 'Are religious beliefs "enabling mechanisms for survival"?' Zygon 15:193-201.

Axelrod R. 1984. The Evolution of Co-operation (London: Penguin Books), 241 p.

Ayala FJ. 1998. 'Biology precedes, culture transcends: an evolutionist's view of human nature' Zygon 33:507-523.

Bahn P, Flenley J. 1992. Easter Island, Earth Island: a message from our past for the future of our planet *(London: Thames and Hudson),* 240 p.

Baker B. 1996. 'A reverent approach to the natural world' Bioscience 46(7):475-478.

Ball I, Goodall M, Palmer C, Reader J, editors, 1992. The Earth Beneath: A critical guide to green theology *(London: SPCK),* 216 p.

Barbour IG. 1972. 'Attitudes toward Nature and Technology'. In: Barbour I G, editor. Earth Might be Fair: Reflections on Ethics, Religion and Ecology *(Englewood Cliffs, NJ: Prentice-Hall Inc),* p 146-168.

Barbour IG. 1997. Religion and Science: historical and contemporary issues *(New York: HarperCollins), 368 p.*

Bartholomew DJ. 1984. God of Chance *(London: SCM Press), 181 p.*

Beitzig L. 1992. 'Roman polygyny' Ethology and Sociobiology *13:309-349.*

Berry RJ. 1991. Christianity and the Enviroment: Escapist mysticism or responsible stewardship. *Science and Christian belief 3:3–18.*

Berry RJ. editor, 1993a. Environmental Dilemmas *(London: Chapman & Hall), 271 p.*

Berry RJ. 1993b. 'Green religion and green science'. In: Anon, editor. Explorations in Science and Theology *(London: RSA), p 23-37.*

Berry RJ. 1995. 'Creation and the environment' Science and Christian Belief *7:21-43.*

Berry W. 1993. Christianity and the survival of creation. *Cross Currents 43(2):149–163.*

Best E. 1942. Forest Lore of the Maori *(Wellington: Government Printer), 421 p.*

Birch C, Eakin W, McDaniel J editors. 1990. Liberating life *(Maryknoll, NY: Orbis Books).*

Blackmore S. 1999. The Meme Machine *(Oxford: Oxford University Press).*

Boehm C. 1997. 'Impact of the human egalitarian syndrome on Darwinian selection mechanics'. American Naturalist *150, Supplement:S100-S121.*

Bowker J. 1995. Is God a virus? genes, culture and religion *(London: SPCK), 274 p.*

Boyd R, Richerson. 1985. Culture and the Evolutionary Process *(Chicago: University of Chicago Press), 331 p.*

Boyden S. 1987. Western Civilisation in Biological Perspective *(Oxford: Oxford University Press), 370 p.*

Brennan AA. 1993. 'Environmental decision-making'. In: Berry RJ, editor. Ecological Dilemmas: Ethics and Decisions *(London: Chapman and Hall), p 1-19.*

Brown LR, Flavin C, Postel S. 1990. 'Picturing a sustainable society'. In: Brown LR, editor. State of the World: 1990 *(New York: Norton), p 174.*

Brown RF. 1975. 'On the necessary imperfection of creation: Irenaeus' Adversus Haereses iv,38'. Scottish Journal of Theology *28:17-25.*

Bruntland GH. 1987. Our Common Future *(Oxford: Oxford University Press), 400 p.*

Budiansky S. 1995. Nature's Keepers. The new science of nature management. *London: Weidenfeld & Nicolson), 310 p.*

Burhoe RW. 1970. 'Natural Selection and God' Zygon *7:30-63.*

Burhoe RW. 1979. 'Religion's role in human evolution: the missing link between ape-man's selfish genes and civilized altruism'. Zygon *14:135-162.*

Byrne R. 1995. The Thinking Ape: Evolutionary Origins of Intelligence *(Oxford: Oxford University Press), 266 p.*

Campbell DT. 1975. 'The conflict between social and biological evolution and the concept of orginal sin'. Zygon *10:234-249.*

Capon RF. 1983. Parables of Grace *(Grand Rapids, Mich.: WB Eerdmans), 184 p.*

Capon RF. 1996. The Astonished Heart *(Grand Rapids, Mich.: WB Eerdmans), 122 p.*

Caughley G. 1983. The Deer Wars *(Auckland: Heinemann Publishers), 187 p.*

Caughley G, Gunn A. 1996. Conservation Biology in Theory and Practice *(Cambridge, Mass: Blackwell Science), 459 p.*

Cavanaugh M. 1996. Biotheology: A new synthesis of Science and Religion *(Lanham, MD: University Press of America), 334 p.*

Chial DL. 1996. 'The ecological crisis: a survey of the WCC's recent responses'. Ecumenical Review *48:53-61.*

Chomsky N, Barsamian D. 1998. The Common Good. Monroe *(ME: Common Courage Press), 190 p.*

Colinvaux P. 1980. The Fates of Nations: A Biological Theory of History *(London: Penguin), 269 p.*

Crosby AW. 1986. Ecological Imperialism: The Biological Expansion of Europe, *900-1900 (Cambridge: Cambridge University Press), 368 p.*

Cupitt D. 1984. The Sea of Faith *(London: BBC), 286 p.*

Daly HE, Cobb J. 1990. For the Common Good: redirecting the economy towards community, the environment and a sustainable future *(London: Green Print Merlin Press), 482 p.*

Daly M, Wilson M. 1997. 'Cinderella revisited'. In: Beitzig L, editor. Human Nature: a critical reader *(New York: Oxford University Press), p 172-4.*

Darwin C. 1871. The Descent of Man. *1981 reprint (Princeton: Princeton University Press).*

Dawkins R. 1989. The Selfish Gene *(Oxford: Oxford University Press), 352 p.*

Dawkins R. 1996. Climbing Mount Improbable *(Penguin).*

de Waal F. 1982. Chimpanzee Politics *(Baltimore: Johns Hopkins Press), 227 p.*

de Waal F. 1989. Peacemaking Among Primates *(Cambridge Mass: Harvard University Press), 294 p.*

de Waal F. 1996. Good Natured: The Origins of Right and Wrong in Humans and Other Animals *(Cambridge, Mass: Harvard University Press), 296 p.*

de Waal F. 2000. 'Primates—a natural heritage of conflict resolution'. Science *289(28 July):586-590.*

Dennett DC. 1995. Darwin's Dangerous Idea *(New York: Simon and Schuster), 586 p.*

Dugatkin LA, Mesterton-Gibbons M, Houston A I. 1992. 'Beyond the Prisoner's Dilemma: toward models to discriminate among mechanisms of co-operation in nature'. Trends in Ecology and Evolution *7(6):202-205.*

Dugatkin LA, Wilson DS, Farrand L III, Wilkens RT. 1994. 'Altruism, tit for tat and 'outlaw' genes'. Evolutionary Ecology *8:431-437.*

Durham W. 1991. Co-evolution: genes, culture and human diversity *(Stanford, CA: Stanford University Press).*

Duyker E. 1994. An Officer of the Blue: Marc-Joseph Marion Dufresne, South Sea Explorer, 1724-72 *(Melbourne: Melbourne University Press), 229 p.*

Ecologist T. 1993. Whose Common Future? *(London: Earthscan Publications), 216 p.*

Ehrenfeld D. 1981. The Arrogance of Humanism *(New York: Oxford University Press), 286 p.*

Ehrlich PR. 1997. A World of Wounds: Ecologists and the Human Dilemma *(Ecology Institute: Excellence in Ecology)*, 210 p.

Ehrlich PR, Ehrlich A E. 1996. Betrayal of Science and Reason: How anti-environmental rhetoric threatens our future *(Washington DC: Island Press)*, 335 p.

Eldredge N. 1995. Reinventing Darwin: the great evolutionary debate *(London: Weidenfeld and Nicolson)*, 244 p.

Elliott H. 1997. 'A general statement of the Tragedy of the Commons'. Population and Environment 18(6):515-531.

Flannery T. 1994. The Future Eaters *(Port Melbourne: Reed Books)*, 423 p.

Fleming CA. 1962. History of the New Zealand land bird fauna. Notornis 9:270-274.

Foley RA. 1996. 'An evolutionary and chronological framework for human social behaviour'. In Runciman W, Maynard Smith J, Dunbar R, editors. Evolution of social behaviour patterns in primates and man *(Oxford: The British Academy and Oxford University Press)*, p 297.

Friends of the Earth Europe. 1995. Towards Sustainable Europe *(Brussels: Friends of the Earth Europe)*.

Galbraith JK. 1992. The Culture of Contentment *(London: Penguin)*, 195 p.

Galbraith JK. 1996. The Good Society *(London: Reed International Books)*, 152 p.

Galbreath R. 1989. Walter Buller, The Reluctant Conservationist *(Wellington: GP books)*, 336 p.

Galleni L. 1992. 'Relationships between scientific analysis and the world view of Pierre Teilhard de Chardin'. Zygon 27:153-166.

Galvin R, Kearns R, editors. 1989. Repainting the Rainbow: ecology and Christian living *(Auckland: Christian Ecology Group and Maclaurin Chaplaincy, University of Auckland)*, 70 p.

Gerle E. 1995. 'In search of a global ethics: theological, political and feminist perspectives based on a critical analysis of JPIC and WOMP'. Lund Studies in Ethics and Theology 2:1-273.

Gosling D. 1992. A New Earth: Covenanting for Justice, Peace and the Integrity of Creation *(London: Council of Churches for Britain and Ireland (CCBI))*, 108 p.

Gould SJ. 1977. Ever Since Darwin *(New York: WW Norton)*, 285 p.

Gould SJ. 1983a. 'Hen's Teeth and Horse's Toes'. In Gould S J, editor. Hen's Teeth and Horse's Toes *(New York: WW Norton)*, p 177-186.

Gould SJ. 1983b. 'Nonmoral Nature'. In Gould S J, editor. Hen's Teeth and Horse's Toes *(New York: WW Norton)*, p 32-45.

Gould SJ. 1993a. 'An earful of jaw'. In Gould S J, editor. Eight Little Piggies *(London: Penguin)*, p 95-108.

Gould SJ. 1993b. 'Eight little piggies'. In Gould S J, editor. Eight Little Piggies *(London: Penguin)*, p 63-78.

Gould SJ. 1993c. 'Full of hot air'. In Gould S J, editor. Eight Little Piggies *(London: Penguin)*, p 109-120.

Granberg-Michaelson W. 1994. 'Creation in ecumenical theology'. In Hallman D G, *editor.* Ecotheology: voices from North and South *(Geneva: WCC Publications), p 96-106.*

Granberg-Michaelson W. 1992. Redeeming the Creation: the Rio Earth Summit-challenge for the churches *(Geneva: WCC Publications), 90 p.*

Grant C. 1986. 'The gregarious metaphor of the selfish gene'. Religious Studies 27:431-450.

Grant C. 1993. 'The odds against altruism: the sociobiology agenda'. Perspectives on Science and Christian Faith 45(2):96-110.

Grant-Taylor D, O'Shaughnessy B. 1992. Rotorua Geothermal Field: a review of the field response to closure *(Whakatane: Bay of Plenty Regional Council), Report nr 7: 1-57.*

Grove–White R, O'Donovan O. 1996. Analternative approach. *In: Attfield R, Dell K, editors.* Values, Conflict and the Enviroment. Second ed. *(Aldershot: Avebury Press.) p. 117–133*

Hall JD. 1986. Imaging God: Dominion as Stewardship *(Grand Rapids: WB Eerdmans Publishing Co), 248 p.*

Hallman DG. 1994. 'Beyond "North-South" Dialogue'. *In Hallman D G, editor.* Ecotheology: Voices from North and South *(Geneva: WCC Publications), p 3-9.*

Hardin G. 1968. 'The Tragedy of the Commons'. Science 162:1243-1248.

Hardin G. 1993. Living within limits: ecology, economics, and population taboos *(Oxford: Oxford University Press), 339 p.*

Hardin G. 1994. 'The tragedy of the unmanaged commons'. Trends in Evolution and Ecology 9(5):199.

Hardin G, Baden J, editors. 1977. Managing the Commons *(San Francisco: Freeman Books).*

Hartley P. 1997. Conservation Strategies for New Zealand *(Wellington: New Zealand Business Roundtable), 526 p.*

Hawken P. 1993. The Ecology of Commerce: a declaration of sustainability *(New York: Harpercollins), 250 p.*

Hay R. 1996. 'Biodiversity Research: the international conservation context'. *In McFadgen B, Simpson P, editors.* Biodiversity. Papers from a seminar series on biodiversity, 14 June to 26 July 1994 *(Wellington: Department of Conservation), p 167-70.*

Hefner P. 1993. The Human Factor: Evolution, Culture and Religion *(Minneapolis: Fortress Press), 317 p.*

Heinen JT, Low RS. 1992. 'Human behavioural ecology and environmental conservation'. Environmental Conservation 19:105-116.

Hick J. 1977. Evil and the God of Love *(San Francisco: HarperCollins), 389 p.*

Honderich T, editor. 1995. The Oxford Companion to Philosophy *(Oxford: Oxford University Press), 1009 p.*

Houghton J. 1997. 'Christians and the Environment: our opportunities and responsibilities'. Science and Christian Belief 9:101-111.

Hughey KFD, Parkes JP. 1996. 'Thar management-planning and consultation under the Wild Animal Control Act' Royal Society of New Zealand, Miscellaneous Series 31:85-90.

Hutchinson GE. 1965. The ecological theatre and the evolutionary play *(New Haven, Conn.: Yale University Press)*.

Irons W. 1996. 'Morality, religion and human evolution'. In Richardson W M, Wildman W J, editors. Religion and Science: History, Method, Dialogue *(New York: Routledge)*.

Irons W. 1997. 'Cultural and biological success'. In Beitzig L, editor. Human Nature: a critical reader *(Oxford: Oxford University Press)*, p 36-49.

Jacobs M. 1996. The Politics of the Real World *(London: Earthscan Publications Ltd)*, 145 p.

John Paul II. 1996. 'Message to the Pontifical Academy of Sciences'. Quarterly Review of Biology 72:381-383.

Jones S. 1996. In the Blood *(London: HarperCollins)*, 302 p.

Kaiser C. 1991. Creation and the History of Science *(Grand Rapids, Michigan: WB Eerdmans)*, 316 p.

Kaiser CB. 1993.' The creationist tradition in the history of science'. Perspectives on Science and Christian Faith 45:80-89.

Kaiser CB. 1996. 'The integrity of creation and the social nature of God'. Scottish Journal of Theology 49:261-290.

Kee A. 1982. Constantine versus Christ: the Triumph of Ideology *(London: SCM Press)*, 182 p.

Kelsey J. 1997. The New Zealand Experiment: a world model for structural adjustment? *(Auckland: Auckland University Press)*, 433 p.

Keohane R, Ostrom E, editors. 1995. Local Commons and Global Interdependence: Heterogeneity and Cooperation in two Domains *(London: Sage Publications)*, 261 p.

King CM. 1984. Immigrant Killers: Introduced Predators and the Conservation of Birds in New Zealand *(Auckland: Oxford University Press)*, 224 p.

King CM. 1990. The Handbook of New Zealand Mammals *(Auckland: Oxford University Press)*, 600 p.

King CM. 1997. 'Is Theology Useful?' Science and Spirit 8(2):6-7.

King CM. 2001. 'Ecotheology: a marriage between secular ecological science and rational, compassionate faith'. Ecotheology 10:40-69.

Kühn T. 1970. The Structure of Scientific Revolutions *(Chicago: University of Chicago Press)*, 210 p.

Küng H. 1989. 'Paradigm change in theology: a proposal for discussion'. In Küng H, Tracey D, editors. Paradigm change in theology *(Edinburgh: T&T Clark Ltd)*, p 3-33.

Küng H. 1990. Global Responsibility: In search of a new world ethic *(London: SCM Press),*

Larson EJ, Witham L. 1997. 'Scientists are still keeping the faith'. Nature *386:435-436.*

Lawton JH. 1994. 'What will you give up?' Oikos *71(3):353-354.*

Leakey R, Lewin R. 1996. The Sixth Extinction: Biodiversity and its Survival *(London: Weidenfeld and Nicolson), 271 p.*

Leopold A. 1949. A Sand County Almanac *(New York: Oxford University Press), 228 p.*

Lewis CS. 1942. The Screwtape Letters *(London: Geoffrey Bles),*

Lewis CS. 1943. Out of the Silent Planet *(London: Pan),*

Lewis CS. 1971. Man or Rabbit? Undeceptions: Essays on Theology and Ethics *(London: Geoffrey Bles)*

Lewis CS. 1977. The Pilgrim's Regress *(London: Collins),*

Lewontin RC. 1991. The Doctrine of DNA: Biology as Ideology *(London: Penguin),*

Lindberg DC. 1992. The Beginnings of Western Science: The European Scientific Tradition in Philosophical, Religious and Institutional Context, 600 BC to AD 1450 *(Chicago: University of Chicago Press), 455 p.*

Lucas JR. 1979. 'Wilberforce and Huxley: a legendary encounter'. The Historical Journal *22:313-330.*

Macdonald DW. 1983. 'The ecology of carnivore social behaviour'. Nature *301:379-384.*

Martin D. 1997. Does Christianity cause war? *(Oxford: Clarendon Press), 226 p.*

Maynard Smith J, Szathmary E. 1995. The Major Transitions in Evolution *(Oxford: WH Freeman), 346 p.*

McCay BM, Acheson JM, editors. 1987. The Question of the Commons: the culture and ecology of communal resources *(Tucson: University of Arizona Press), 439 p.*

McDaniel JB. 1989. Of God and Pelicans: a theology of reverence for life *(Louisville, Kentucky: Westminster/John Knox Press), 168 p.*

McDaniel JB. 1995. With Roots and Wings: Christianity in an age of ecology and dialogue *(Maryknoll, New York: Orbis Books), 243 p.*

McDonagh S. 1994. Passion for the Earth: the Christian vocation to promote justice, peace and the integrity of creation *(London: Geoffrey Chapman), 164 p.*

McFague S. 1993. The Body of God: an ecological theology *(London: SCM Press), 274 p.*

McGrath AE. 1994. Christian Theology: An Introduction *(Oxford: Blackwell), 510 p.*

McGrath AE. 1998. The Foundations of Dialogue in Science and Religion *(Malden, Mass.: Blackwell Publishers Inc), 256 p.*

McManners J. 1992. The Oxford Illustrated History of Christianity *(Oxford: Oxford University Press), 724 p.*

Mesle CR. 1993. Process Theology: A Basic Introduction *(St Louis, Missouri: Chalice Press), 148 p.*

Midgley M. 1978. Beast and Man: The Roots of Human Nature *(London: Methuen University Paperback), 377 p.*

Midgley M. 1979. 'Gene-juggling'. Philosophy *54:439-458.*

Midgley M. 1994. The Ethical Primate: Humans, Freedom and Morality *(London: Routledge), 193 p.*

Moltmann J. 1985. God in Creation. Kohl M, translator *(London: SCM Press), 365 p.*

Moore NW. 1987. The Bird of Time: the science and politics of nature conservation *(Cambridge: Cambridge University Press), 290 p.*

Morton J. 1989. Christ, Creation and the Environment *(Auckland: Anglican Communications, for the General Synod of the Church of the Province of New Zealand), 84 p.*

Murphy N. 1990. Theology in the Age of Scientific Reasoning *(Ithaca, New York: Cornell University Press), 215 p.*

Murphy N, Ellis GFR. 1996. On the Moral Nature of the Universe: theology, cosmology, and ethics *(Minneapolis: Fortress Press), 268 p.*

Newbigin L. 1986. Foolishness to the Greeks: the Gospel and Modern Culture *(Grand Rapids, Michigan: W B Eerdmans Publishing Co), 156 p.*

Niles DP, editor. 1992. Between the Flood and the Rainbow: Interpreting the Conciliar Process of Mutual Commitment (Covenant) to Justice, Peace and the Integrity of Creation *(Geneva: WCC Publications), 192 p.*

Niles P. 1989. Resisting the Threats to Life: Covenanting for Justice, Peace and the Integrity of Creation *(Geneva: Risk Books, World Council of Churches), 85 p.*

Northcott MS. 1994. 'Is there a green Christian ethic?' Studies in Christian Ethics *7:32-51.*

Oelschlaeger M. 1994. Caring for Creation:an Ecumenical Approach to the Environmental Crisis *(Newhaven Conn: Yale University Press), 285 p.*

Ormerod P. 1994. The Death of Economics *(London: faber and faber), 230 p.*

Oye KA, Maxwell J H. 1995. 'Self interest and environmental management'. In Keohane *RO, Ostrom E, editors.* Local Commons and Global Interdependence: heterogeneity and cooperation in two domains *(London: Sage Publications), 191-221. p.*

Page R. 1996. God and the Web of Creation *(London: SCM Press), 188 p.*

Pagels E. 1988. Adam, Eve and the Serpent *(London: Penguin), 189 p.*

Palmer G. 1995. Environment: The International Challenge *(Wellington: Victoria University Press), 191 p.*

Passmore J. 1980. Man's Responsibility for Nature *(London: Duckworth), 227 p.*

Peacocke A. 1979. Creation and the world of science *(Oxford: Clarendon Press), 389 p.*

Peacocke A. 1993. Theology for a Scientific Age *(London: SCM Press), 438 p.*

Peacocke A. 1995. 'The challenge of science to the thinking church'. Modern Believing *36(4):15-26.*

Peacocke A, Hodgson P. 1996. The Judaeo–Christian Tradition. *In: Attfield R, Dell K, editors. Values, Conflict and the Enviroment. Second ed. (Aldershot: Avebury Press.) p. 141–146*

Poole M. 1994. *'A critique of aspects of the philosophy and theology of Richard Dawkins'.* Science and Christian Belief 6:41-59.

Primavesi A. 1991. From Apocalypse to Genesis *(Tunbridge Wells, Kent: Burns and Oates), 324 p.*

Quammen D. 1996. The Song of the Dodo: Island biogeography in an age of extinctions *(London: Hutchinson), 702 p.*

Rainbow S. 1993. Green Politics *(Auckland: Oxford University Press), 106 p.*

Rasmussen L. 1996. Earth Community, Earth Ethics *(Geneva: WCC Publications), 366 p.*

Reynolds V, Tanner R. 1995. The Social Ecology of Religion *(Oxford: Oxford University Press), 322 p.*

Richards D. 1991. *'Post-communal land ownership: poverty and political philosophy'. In Andelson RV, editor.* Commons Without Tragedy: Protecting the Environment from Overpopulation-a new approach *(London: Shepheard-Walwyn), p 83-108.*

Ridley M. 1993. The Red Queen: Sex and the Evolution of Human Nature *(London: Penguin), 404 p.*

Ridley M. 1996. The Origins of Virtue *(London: Viking Press), 295 p.*

Rolston III H. 1992. *'Disvalues in nature'.* The Monist 75:250-278.

Rolston III H. 1994. *'Does nature need to be redeemed?'* Zygon 29:219-229.

Rue L. 1998. *'Sociobiology and moral discourse'.* Zygon 33:525-533.

Rue LD. 1989. Amythia: Crisis in the Natural History of Western Culture *(Tuscaloosa, Alabama: University of Alabama Press), 206 p.*

Rue LD. 1994. By the Grace of Guile: The Role of Deception in Natural History and Human Affairs *(New York: Oxford University Press), 359 p.*

Runciman WG, Maynard Smith J, Dunbar R I M, editors. 1996. Evolution of social behaviour patterns in primates and man *(Oxford: Oxford University Press, for The British Academy), 297 p.*

Ruse M. 1986. Taking Darwin Seriously: A Naturalistic Approach to Philosophy *(Oxford: Basil Blackwell), 303 p.*

Santmire P. 1985. The Travail of Nature: the ambiguous ecological promise of Christian theology *(Minneapolis: Fortress Press), 274 p.*

Schmitz-Moormann K. 1995. *'The future of Teilhardian theology'.* Zygon 30:117-129.

Sheehan T. 1988. The First Coming: How the Kingdom of God became Christianity. *(New York: Vintage Books (Random House)), 287 p.*

Simkins RA. 1994. Creator and Creation: Nature in the Worldview of Ancient Israel *(Peabody, Mass. USA: Hendrickson Publishers Inc), 306 p.*

Singer P. 1983. The Expanding Circle: ethics and sociobiology *(Oxford: Oxford University Press), 190 p.*

Smith A. 1776. An Enquiry into the Nature and Causes of the Wealth of Nations *(New York: Modern Library) (1937 reprint)*.

Stauffer RC. 1975. Charles Darwin's Natural Selection *(Cambridge: Cambridge University Press)*, 382 p.

Stevenson GG. 1991. Common Property Economics: a general theory and land use applications *(Cambridge: Cambridge University Press)*,

Swadling H. 1989. *'The practical implications of the ecological crisis'. In Galvin R, Kearns R, editors.* Repainting the Rainbow: ecology and Christian living. *(Auckland: Christian Ecology Group and Maclaurin Chaplaincy, University of Auckland)*, p 26-34.

Sylvan R. 1992. On the value core of deep–green theory. *In: Oddie G, Perrett RW, editors. Justice, Ethics and New Zealand Society. Auckland: Oxford University Press.* p 222–229

Theissen G. 1984. Biblical Faith: An Evolutionary Approach. *Bowden J, translator (London: SCM Press)*, 194 p.

Thomas K. 1983. Man and the Natural World *(Harmondsworth: Penguin Books)*, 426 p.

Thomson D. 1991. Selfish Generations? The Ageing of New Zealand's Welfare State *(Wellington: Bridget Williams Books)*, 233 p.

Turner H. 1998. The roots of science: an investigative journey through the world's religions *(Auckland: The Deepsight Trust)*, 204 p.

van Till HJ. 1996. *'Basil, Augustine, and the Doctrine of Creation's Functional Integrity'.* Science and Christian Belief *8:21-38*.

Vink N, Kassier W E. 1987. *'The "tragedy of the commons" and livestock farming in southern Africa'.* South African Journal of Economics *55:165-182*.

Vitousek P, Mooney H A, Lubchenco J, Melillo J M. 1997. *'Human domination of Earth's ecosystems'.* Science *277:494-499*.

Wade R. 1987. *'The management of common-property resources: collective action as an alternative to privatisation or state regulation'.* Cambridge Journal of Economics *11:95-106*.

Ward K. 1992. Defending the Soul *(Oxford: Oneworld (Hodder and Stoughton))*, 175 p.

Ward K. 1996. God, Chance and Necessity *(Oxford: Oneworld)*, 212 p.

White L. 1967. *'The historic roots of our ecologic crisis'.* Science *155(3767):1204-1207*.

Whitehead AN. 1927. Science and the Modern World *(Cambridge: Cambridge University Press)*.

Whiten A, Byrne RW, editors. 1997. Machiavellian Intelligence II: extensions and evaluations *(Cambridge: Cambridge University Press)*, 403 p.

Wilkinson L. 1993. *'Christianity and the Environment: Reflections on Rio and Au Sable'.* Science and Christian Belief *5:139-145*.

Willey K. 1979. When the Sky Fell Down: The Destruction of the Tribes of the Sydney Region, 1788-1850s *(Sydney: Collins)*, 231 p.

Williams GC. 1966. Adaptation and natural selection. A critique of some current evolutionary thought *(Princeton: Princeton University Press), 307 p.*

Williams GC. 1992. Natural Selection: Domains, Levels and Challenges *(New York: Oxford University Press), 208 p.*

Williams GC. 1996. Plan and Purpose in Nature *(London: Weidenfeld & Nicolson), 191 p.*

Williams PA. 1998. 'Evolution, Sociobiology and the Atonement'. Zygon 33:557-570.

Williamson R. 1992. '"What God has joined together, let no one put asunder": Reflections on JPIC at the Canberra Assembly'. In Niles D, editor. Between the Flood and the Rainbow: Interpreting the Conciliar Process of Mutual Commitment (Covenant) to Justice, Peace and the Integrity of Creation *(Geneva: WCC Publications), p 82-101.*

Wilson D. 1992. 'On the relationship between evolutionary and psychological definitions of altruism and selfishness'. Biology and Philosophy 7:61-68.

Wilson D. 1997a. 'Altruism and organism: disentangling the themes of multi-level selection theory'. American Naturalist 150, Supplement:S122-S134.

Wilson D. 1997b. 'Introduction: multi-level selection theory comes of age'. American Naturalist 150 Supplement:S1-S4.

Wilson D, Sober E. 1994. 'Reintroducing group selection to the human behavioural sciences'. Behavioural and Brain Sciences 17:585-654.

Wilson EO. 1975. Sociobiology: the new synthesis *(Cambridge, Mass.: Harvard University Press), 697 p.*

Wilson EO. 1977. On Human Nature *(Cambridge, Mass: Harvard University Press).*

Wilson EO. 1998. Consilience: The unity of knowledge *(London: Little, Brown Co), 374 p.*

Wilson R. 1982. From Manapouri to Aramoana: the battle for New Zealand's environment *(Auckland: Earthworks Press), 192 p.*

Wolpert L. 1992. The Unnatural Nature of Science *(London: faber and faber), 191 p.*

Wrangham R, Peterson D. 1996. Demonic Males: Apes and the origin of human violence *(London: Bloomsbury), 350 p.*

Wright LW. 1980. 'Decision making and the logging industry:an example from New Zealand'. Biological Conservation 18:101-115.

Wright R. 1994. The Moral Animal *(New York: Random House), 466 p.*

Wyman RL, Steadman DW, Sullivan ME, Walters-Wyman M F. 1991. 'Now what do we do?' In Wyman R L, editor. Global Climate Change and Life on Earth. *(New York: Routledge, Chapman & Hall), p 252-263.*

Zacks R. 1997. 'What are they thinking? Student's reasons for rejecting evolution go beyond the Bible'. Scientific American 277(October 1997):34.

Index